Holding On to Reality

Holding On to *Reality*

The Nature of
Information
at the Turn of
the Millennium

Albert Borgmann

The University of Chicago Press
Chicago and London

ALBERT BORGMANN has taught philosophy at the University of Montana since 1970 and specializes in the philosophy of society and culture. He is the author of *Technology and the Character of Contemporary Life* (1984) and *Crossing the Postmodern Divide* (1992).

The University of Chicago Press, Chicago 60637
The University of Chicago Press, Ltd., London
© 1999 by The University of Chicago
All rights reserved. Published 1999
08 07 06 05 04 03 02 01 00 99   1 2 3 4 5

ISBN: 0-226-06625-8 (cloth)

Library of Congress Cataloging-in-Publication Data

Borgmann, Albert.
    Holding on to reality : the nature of information at the turn of
the millennium / Albert Borgmann.
        p.   cm.
    Includes bibliographical references (p.   ) and index.
    ISBN 0-226-06625-8 (acid-free paper)
    1. Communication.—Philosophy. 2. Information theory. I.
Title.
P90.B638   1999
302.2—dc21   98–38578
                                                            CIP

♾ The paper used in this publication meets the minimum requirements
of the American National Standard for Information Sciences—
Permanence of Paper for Printed Library Materials, ANSI Z39.48-1992.

For Andrea, Caitlin, and Kendra

Contents

Information vs. Reality

Information can illuminate, transform, or displace reality. When failing health or a power failure deprives you of information, the world closes in on you; it becomes dark and oppressive. Without *information about reality,* without reports and records, the reach of experience quickly trails off into the shadows of ignorance and forgetfulness.

In addition to the information that discloses what is distant in space and remote in time, there is information that allows us to transform reality and make it richer materially and morally. As a report is the paradigm of information about reality, so a recipe is the model of *information for reality,* instruction for making bread or wine or French onion soup. Similarly there are plans, scores, and constitutions, information for erecting buildings, making music, and ordering society.

Information about reality exhibits its pristine form in a natural setting. An expanse of smooth gravel is a sign that you are close to a river. Cottonwoods tell you where the river bank is. An assembly of twigs in a tree points to ospreys. The presence of ospreys shows that there are trout in the river. In the original economy of signs, one thing refers to another in a settled order of reference and presence. A gravel bar seen from a distance refers you to the river. It is a sign. When you have reached and begun to walk on the smooth and colored stones, the gravel has become present in its own right. It is a thing. And so with the trees, the nest, the raptors, and the fish.

While natural signs emerge from their environment and disappear in it again, conventional signs have an unnatural prominence and stability. Stones that are piled up in a cairn show a concentration and an angle of repose that set them apart from their surroundings. Conventional signs become truly distinctive vehicles of information when they not only stand out from nature the way cairns do, but are also detached from their environment and rendered mobile as first happened with notches on sticks and pebbles in pockets, and then with clay tokens in pouches, marks on clay tablets, letters on papyrus, and maps on parchment. Signs came to stand apart from things and at the origin of entirely new things. Covenants helped tribes to become nations, plans guided the construction of cathedrals, and scores enabled musicians to perform cantatas. An economy of cultural signs came to enrich the realm of natural signs.

This picture of a world that is perspicuous through natural information and prosperous through cultural information has never been more than a norm or a dream. It is certainly unrecognizable today when the paradigmatic carrier of information is neither a natural thing nor a cultural text, but a technological device, a stream of electrons conveying bits of information. In the succession of natural, cultural, and technological information, both of the succeeding kinds heighten the function of their predecessor and introduce a new function. Cultural information through records, reports, maps, and charts discloses reality much more widely and incisively than natural signs ever could have done. But cultural signs also and characteristically provide information for the reordering and enriching of reality. Likewise technological information lifts both the illumination and the transformation of reality to another level of lucidity and power. But it also introduces a new kind of information. To information *about* and *for* reality it adds *information as reality.* The paradigms of report and recipe are succeeded by the paradigm of the recording. The technological information on a compact disc is so detailed and controlled that it addresses us virtually *as* reality. What comes from a recording of a Bach cantata on a CD is not a report about the cantata nor a recipe—the score—for performing the cantata, it is in the common understanding music itself. Information through the power of technology steps forward as a rival of reality.

Today the three kinds of information are layered over one another in one place, grind against each other in a second place, and are heaved and folded up in a third. But clearly technological information is the most prominent layer of the contemporary cultural landscape, and increasingly it is more of a flood than a layer, a deluge that threatens to erode, suspend, and dissolve its predecessors.

As a consequence, our world abounds with information. You wake up to the news on the radio, read the paper for breakfast, are immersed in signs as you make your way to the office, sit down to fire up your computer—that really opens the floodgate of information—return home, turn on the television set and let waves of information wash over you until you go to bed. Especially in the form of advertising, information, as Brent Staples has remarked, "is rapidly expanding to fill every salable space—which is to say, every space that's empty." It has profaned even the sacred precinct of Yankee Stadium's baseball diamond. "Think now," says Staples, "of a world devoid of quiet and empty, where every surface shouts and every silence is filled."[1]

The roar of information continues to rise, fed by prodigious advances in information technology. When our culture assumes its official voice to pronounce summarily on the effect of the new information technology, it recites a well-worn formula, telling us that this technology will improve "the ways we live, learn, and work."[2] On occasion this studied restraint yields to more enthusiastic claims, and we are told that information is "the wellspring of great fortunes, much as land was a century ago." It will change "the face of the American commercial landscape."[3] And not to leave the issue in doubt, it has been said that fashioning information into intelligent artificial life "is the computer scientist's Great Work as surely as the building of Notre Dame cathedral on the Île-de-France was the Great Work of the medieval artisan."[4]

This enthusiasm is more than a fancy of nerds. Politicians on the extreme right and the far left who can agree on next to nothing are united in their fervor and determination to push ahead with the information highway, and a Congress that habitually delays decisive action or passes crucial bills by narrow margins on 1 February 1996 approved a far-reaching reform of telecommunications by a majority of better than 90 percent in both houses.

Such enthusiasm is not unreasonable. Information technology has already produced an enormous increase in our freedom to select information. There are hundreds of television channels to choose from, millions of people to connect with, oceans of data to seine in an instant, virtual realities to explore and enjoy. The Internet particularly has given many people the liberty to escape the constraints of their age, gender, and race, of their shyness, plumpness, or homeliness, and to set their glamorous inner selves free and adrift on a World Wide Web. Professional people often remark with gratitude how electronic word processing has lowered or entirely removed the barrier they used to, or too often failed to, climb over to get to their writing chores. Theoretically keen and venturesome writers have celebrated the fall of the linear, hierarchical, and austere book and the rise of the flexible and associative multimedia hypertext.[5]

Information technology has become the engine of the postmodern economy. The modern economy was in danger of sclerosis from an excess of mass-produced goods and of chronic if not fatal poisoning due to the toxic conditions it had created in the environment. Information processing has opened up new niches and desires to be filled with customized goods and sophisticated services. It has helped to monitor and clean up the environment and to stretch or recycle resources. Information itself has become a valuable resource and a consumption good that lies easily on the wearied earth.

Yet with all these gains we sometimes feel like the sorcerer's apprentice, unable to contain the powers we have summoned and afraid of drowning in the flood we have loosed. And much like the apprentice we are unable to find the words that would restore calm and order, misspeaking ourselves when we try to get control of our situation.[6] The words that come most easily to the lips of ready critics in our culture concern social justice. Will information technology create a new division between haves and have-nots or deepen the old division? This is surely a fair question. But it tends to divert us from the deeper question of whether the recent and imminent flood of information is good for anybody, rich or poor.

Processing our words with computers and drawing on vast reservoirs of information have rendered our prose prolix and shapeless. But information technology has dissolved more than the contours of our writing. It has infected our sense of identity with doubt and despair. Are my tangible traits just so much noise that distorts the true message of my self? Is my ethereal Internet self the genuine me, freed from the accidents of my place, class, and looks? Or is it a flimsy and truant version of what, for better or worse, I am actually and substantially? Not that the virtual versions of one's self are always the more sublime. In some people the preternatural openness of electronic space ignites firestorms of profanity and hostility that would be unthinkable in face-to-face meetings.

Information is flooding place and property with ambiguity as well. Shadows of doubt and dissatisfaction fall on the belongings that have served you well for decades when weekly a mailing of catalogs urges more elegant and convenient alternatives on you. The daily display of the excitement of the cities and the open spaces of the country makes the place where you live look drab and confined. Both the mooring of one's place and the identity of one's friends get confounded when frequent e-mail makes a distant and unknown person seem closer and more responsive than your friend next door, or when a colleague on the same floor remains cool and distant until he or she begins to open up and confide in you through e-mail.

The farther reaches of reality and the cultural landmarks that used to lend it coherence are being swept off their foundations by information technology. The contents of the National Gallery in London have been transformed into technological information and deposited on a compact disc so that now I can have "the whole National Gallery on my desk."[7] Once digitized, an altar piece can easily be moved from the National Gallery CD to the virtual reality of the church in the Upper Rhine Valley where once it was the center of worship. But the virtual church itself is as free-floating a cultural item as the altar. Whatever is touched by information technology detaches itself from its foundation and retains a bond to its origin that is no more substantial than the Hope diamond's tie to the mine where it was found.

Yet within a global perspective it must seem self-indulgent to worry about these problems. Only about a quarter of the people in this country and 1 percent of the world's population are affluent enough to own a personal computer, have access to a computer network, and need to worry about e-mail flirtations and CD confusions. Yet the less affluent and less educated citizens of the United States are drenched with information as well. Television is the major channel saturating them with news and entertainment. Though they are more passively connected to information, their connection to reality too is profoundly transformed. The breathless glamour of television numbs their ability to confront and endure the gravity and pressure of reality. Information is the element of technological affluence that invades the culture of poor and premodern countries most quickly and easily. First come the transistor radios and then the television sets, the latter few in number but watched by many. If information is not the medium of an overwhelmingly new culture, it is at least the entering wedge that permits indigenous cultures to seep away and disappear.

For a millennium after the birth of Christ, so the Book of Revelation tells us, the devil was to be bound and thrown into a pit so "that he should deceive the nations no more. . . . And when the thousand years are expired, Satan shall be loosed out of his prison, and shall go out to deceive the nations." But at length, Satan is defeated, and the seer of Revelation sees "a new heaven and a new earth."[8] All this failed to happen in the year 1000 and will not come to pass in the year 2000 either unless W. B. Yeats is right and we will see

> That twenty centuries of stony sleep
> Were vexed to nightmare by a rocking cradle,
> And what rough beast, its hour come round at last,
> Slouches towards Bethlehem to be born?[9]

So far, however, the word millennium has retained the sense of both renewal and crisis. Social critics and information theorists are divided on whether information is the devil or the Second Coming. Surely the answer is not one or the other. Not to be mired in endless and inconclusive qualifications, however, we need both a theory and an ethics of information—a theory to illuminate the structure of information and an ethics to get the moral of its development. My hope is that the theory will lend perspicuity to the ethics, that the ethics will give the theory some force, and that, once we have understood information, we will see that the good life requires an adjustment among the three kinds of information and a balance of signs and things.

Part One

Natural Information
Information about Reality

Chapter One

The Decline of Meaning and the
Rise of Information

The Origin of Information

"Information" is an old word, and the verb "to inform" is older yet. Latin *informare*, as Cicero (106–43 B.C.E.) used it, meant to impose a form on some thing, particularly on the mind, in order to instruct and improve it. For the medieval scholastics, information was the companion of materialization. They thought of things as consisting of form and matter, the form informing matter, matter materializing the form. Our word "information" was born in the late Middle Ages when English itself emerged as the prodigious child of the Germanic and the Romance. Information led an inconspicuous semantic life until in the second half of the twentieth century it gradually and at first hesitantly moved to center stage.

The birth certificate of information as a prominent word and notion is an article published in 1948 by Claude Shannon. "Information," however, does not figure in the title, "The Mathematical Theory of Communication."[1] In 1960, Fritz Machlup published *The Production and Distribution of Knowledge in the United States*, a title that sounds odd today. It is *information*, we would think, that can be produced and distributed. The thought did occur to Machlup. He acknowledged the rise of information and considered its inclusion in his title, but in the end he clung to the tradition that began with Plato wherein the grand and explicit topic is knowledge while information is dimly perceived trouble.[2]

Our notion of *information* was conceived as an answer to problems that were born at just about the time the word "information" was. Like all premodern cultures, the late medieval world was alive with meaning. Holy things proclaimed God; nature was pervaded by benign, seductive, and perilous spirits. Reality spoke powerfully to people; it was overwhelmingly significant and eloquent.

At the beginning of the modern era, however, a semantic ice age began in western Europe and has since enveloped the globe. The Greek legend of Pygmalion and Mozart's *Don Giovanni* have scenes of life infusing stone—Venus animating Pygmalion's lovely statue to gladden his heart and God enlivening the grim statue of the commendatore to bring Don Giovanni to judgment. These stories have inspired Romantic painters and contemporary set designers to give us pictures of eloquence flowing into lifeless matter. The reverse has happened in the modern period. Eloquence and meaning began to drain from reality.

A historical event of such magnitude has antecedents and consequences—the collapse of medieval culture, the rise of science and technology. It is not governed, however, by a cosmic law or a universal pattern. It is in its entirety no one's fault or accomplishment. But it does include and engage thinkers, doers, and makers. Thus philosophers since the beginning of the modern period have noted, analyzed, and perhaps accelerated the loss of significant structure. They thought of their culture as being in the position of the citizens of Venice who one fine day in 1902 were contemplating a heap of bricks where for 750 years the Campanile of San Marco had stood. Overnight the major landmark of civic and religious life had collapsed.[3]

Some observers of contemporary culture accuse science of having willfully and maliciously destroyed the spiritual and moral edifices that have served us for so long.[4] Most philosophers, to the contrary, regard this collapse with cool detachment. When the lightning of scientific inquiry struck, so they see it, our cultural constructions simply failed to hold.

It took the citizens of Venice five years to decide whether to rebuild the Campanile and how. But rebuild it they did. Philosophers since the seventeenth century have attempted to reconstruct meaning on the commonly accepted foundation of scientific reality. But to this day there is no widely shared view of how to get from scientific to significant structure, from atoms and molecules to patterns or landmarks that would order and guide our lives.[5]

Structural Information

Something seems missing in a world consisting only of matter and energy—some principle of order or structure—and information appears to be the needed ingredient.[6] Information, says Donald Mac-Kay, is "that which determines form," either by constructing or by selecting it.[7] Science has shown that reality is structured all the way down. At bottom it consists not of "atomless gunk," or perhaps four types of gunk—earth, water, air, and fire—but rather of a finite number of definite particles, lawfully related one to the other.[8] Hence comes the hope that we might build up substantial information about our world from the elementary information we find at the bottom of reality.

Edward Fredkin has pushed this idea farthest. He has us imagine the universe as consisting of a space lattice of cells, a grand crystalline structure of cubes, where each cell or cube can be in one of a finite number of states. The structure comes to life according to a few relatively simple rules that tell the entire structure to pulse from state to state and tell each cell how to change from one state to the next. Simplicity begets and illuminates complexity. All we need to do is uncover the digital rules that both underlie and transcend particle physics.[9]

I will call information that is equated with structure *structural information* and leave it to physicists to determine whether it is a notion that is helpful for their work.[10] When used to illuminate contemporary culture, *structural information,* like *text* in deconstructionism, amounts to a distinction between everything and nothing. Everything is information, nothing is not. The purpose of such distinctions is to push unclean stuff into the abyss of nothingness—no more confusions for the information theorist, no more objective things for the deconstructionist.

But barrenness is the price of cleanliness. A concept is helpful only when it enables us to make distinctions within reality, not when it levels all distinctions and reduces everything to texts or information. Is a tire information? "There are professors of physics," says Robert Wright, "who would pat the tire warmly and say, 'That's a lot of information you are looking at there.'" But such "physicist's information," as Wright calls it, is not what we need. "We're interested," he says, "in 'real-life information'—information used by organisms (including us), whether internally or externally."[11]

Cognitive Information

As Wright suggests, we can hope to separate helpful from general information by making cognition the principle of selection and by attending only to the information that is taken up by a plant, an animal, a computer perhaps, and most especially by a human being. I call what has so been selected *cognitive information*.

There is no doubt that the investigation of how the structures of the environment mesh and fail to mesh with those of ears, eyes, and brains has been illuminating. Cognitive information is a fruitful focus for physiology, psychology, and cognitive science.[12] In social and cultural theory, it has been an unsettling force to some and a liberating one to others. Once awareness has become the measure of information, it seems to follow that one and the same awareness can be the result of different kinds of external circumstances. As Adelbert Ames put it, "what a person sees when he looks at an object cannot be determined simply from a knowledge of the physical nature of that object, since an unlimited number of different physical objects can give rise to the same perception."[13]

Ames was an ingenious builder of contraptions that engender optical illusions. Most famous is the "Ames room" that, inspected from a peep hole, looks like a room in ordinary perspective but is in fact carefully distorted so that, when persons of equal height are placed in the far corners of the room, one looks like a giant and the other like a dwarf.[14] Ames's work did much to transform the understanding of artists and social theorists about how reality informs the mind.[15] The reactions, first played out in the years after the war, have reverberated into our time and divide between a conservative and a progressive side. To conservative minds, Ames seems to have undermined the notion of an objective reality as a court of last appeal in cases of confusion and disagreement. As progressive observers see it, Ames's work frees human beings from the regimen of reality or, more accurately, from rigid and authoritarian dictates of how everyone must construct his or her world from a reality that of itself allows for indefinitely many constructions.

Yet the suggestion that we should think of information about reality as a construction of reality seems inconsistent, and one's suspicion of Ames's argument is fueled by the fallacy Ames committed when he said that, since different objects can cause the same perception, knowing the object is not to know the perception. Knowing the Ames room, we know what the observer will see. But looking through the peephole, we cannot tell whether we are seeing an Ames room with two ordinary people or an ordinary room with a dwarf in one corner and a giant in the other.

Hence we can reassure conservatives and must disappoint progressives by pointing out that both the ambiguity of reality and our freedom to construct it this way or that evaporate when we place things in context. There is no ambiguity for Ames the experimenter. He knows how things look to the observer and whether her interpretation is correct. The observer's uncertainty in turn is resolved when she enlarges her context by entering the room in question or by inspecting its four sides from the outside. To resolve ambiguity by extending its context has been the answer to skeptics ever since the ancient Greeks made an art of skepticism.

Constructivists and skeptics can raise the stakes, of course, and argue that the context the experimenter sees is itself ambiguous and he is no better off nor any more constrained by reality than his subject and her perception. Every context that is invoked to clear up an ambiguity can in turn be considered ambiguous. But now we face once more a distinction between everything and nothing, and if everything is ambiguous, ambiguity itself becomes nothing. We can call something ambiguous only if it is unambiguously clear what "it," the "something," is, just as we can disagree only if we agree on what it is we are disagreed about. It must be clear that it is that room, seen through this peephole, that we are talking about before we can start to wonder whether *it* is an ordinary or an Ames room we are looking at.

One can, however, conceive of a person's interaction with her environment entirely as a matter of cognitive information without falling into skepticism or constructivism as regards the real world.

To see that this is my mother is to have a certain kind of information. To know that I am now at home is to have another kind. To talk to my mother is to produce information. As Claude Shannon's herald, Warren Weaver, put it, "In oral speech, the information source is the brain, the transmitter is the voice mechanism producing the varying sound pressure (the signal) which is transmitted through the air (the channel)."[16] This approach too can be illuminating, but it disregards the way we speak, and it veils the way we live or perhaps used to live. If someone were to ask me if I had any information about the Rattlesnake Valley in western Montana, I would hesitate to say yes and would instead reply that, well, I *know* the valley, this is where I live. I would give a similar response if I were queried for information about Grete Borgmann.

Information and Presence

But isn't having information the same as knowing? So it is; but there are two kinds of knowledge, direct knowledge and indirect, and only the latter is the same as having information about something or someone. I know *of* Death Valley; I know *that* it is arid and contains the lowest point in the United States. But, I must confess, I do not know *it*. I know *of* Toni Morrison; I know *that* she wrote *Tar Baby* and received the Nobel Prize. But, I regret to say, I do not know *her*. Direct knowing, as it happens, takes a direct object, indirect knowing an indirect object or a dependent clause. French and German have separate verbs for direct (*connaître* and *kennen*) and indirect knowing (*savoir* and *wissen*).

Bertrand Russell marked this distinction as one of *knowledge by acquaintance* vs. *knowledge by description*. His gloss on the distinction suggests that it is of more than linguistic or logical interest and reflects a subtle and crucial issue of manners and morals. Russell explains: "In fact, I think the relation of subject and object which I call acquaintance is simply the converse of the relation of object and subject which constitutes presentation. That is, to say that S has acquaintance with O is essentially the same thing as to say that O is presented to S."[17] The inference we may draw from Russell's point is that if we allow *having information* to obscure the difference between direct and indirect knowing, our sense of the presence of things will become obtuse.[18] A robust sense of reality can easily get lost through technical sophistication. J. L. Austin used to mock the philosophers who punctiliously insisted that "evidence" is needed to support the claim that we are in the presence of tangible reality. It is not the case, says Austin,

> that whenever a "material-object" statement is made, the speaker must have or could produce evidence for it. This may sound plausible enough; but it involves a gross misuse of the notion of "evidence." The situation in which I would properly be said to have *evidence* for the statement that some animal is a pig is that, for example, in which the beast itself is not actually on view, but I can see plenty of pig-like marks on the ground outside its retreat. If I find a few buckets of pig-food, that's a bit more evidence, and the noises and the smell may provide better evidence still. But if the animal then emerges and stands there plainly in view, there is no longer any question of collecting evidence; its coming into view doesn't provide me with more evidence that it's a pig, I can now just *see* that it is, the question is settled.[19]

Something analogous can be said about information. The word normally implies a focal area of nearness and a peripheral realm of farness. "Information," David Israel and John Perry hold, "typically involves a fact [a sign, as I would put it] indicating something about the way things are elsewhere and elsewhen, and this is what makes information useful and interesting."[20]

Austin concludes his instruction for philosophers with an assured "the question is settled." So it may be as regards the notion of evidence or information. But when it comes to leading our lives in contemporary culture, the question of the presence of things and persons is very much open. The leveling of the distinction between direct and indirect knowledge and of the difference between the nearness and farness of reality is not the result of a wrong move in epistemology, but a reflection of the historic decline of meaning. Cultural landmarks, dimensions, and distinctions are dissolving. Everyone is becoming indifferently related to everything and everyone else. This process began with the modern era, and it is now approaching its culmination through information technology. William Mitchell, dean of a school of architecture no less, sees the traditional city, the place of places, being displaced if not replaced by structures of information. Speaking of the "City of Bits," he says: "This will be a city unrooted to any definite spot on the surface of the earth, shaped by connectivity and bandwidth constraints rather than by accessibility and land values, largely asynchronous in its operation, and inhabited by disembodied and fragmented subjects who exist as collections of aliases and agents. Its places will be constructed virtually by software instead of physically from stones and timbers, and they will be connected by logical linkages rather than by doors, passageways, and streets."[21] To judge the significance of this prognosis, we need to know more about the nature of information.

The Nature of Information

The Information Relation

When we look at the world as structural information, we see order and structure everywhere and all the way down. There is nothing that escapes description or explanation. But from the standpoint of structural information, neither is there anything that stands out from the rest with a commanding presence. There is no difference between a fence post and a person. Both are structured through and through. When we assume the stance of cognitive information, we take the distinction between human beings and their world as pivotal. We focus on those structures of information that connect the former with the latter. But once again, there is something that remains indistinct. What lies invisibly within the cognitive compass is the distinction between things that are present immediately and things we have heard or read about. Thus a still more distinctive sense of information is needed. We may call it *instructive information* because in its original and central sense it teaches us about what is remote in space or time.

Instructive information is not a carrier that brings a distant thing to our doorstep. It is not true that a sign *is* a distant thing; it rather is *about* a distant thing. What it delivers is not a thing but the sense of a thing—a message. For information to work that way, there have to be signs, objects of some sort that are about some thing, objects whose function is reference rather than presence.

Thus an ecology of things and an economy of signs are crucial to information and to life in a world that is both engaging and perspicuous. If the world consisted of signs only, nothing would be present and address us in its own right. A sign is the promise of some thing. In a world exclusively of signs, one promise would proclaim another, but nothing would ever be fulfilled. If the world consisted of things only, we would be confined to our immediacy. Nothing could refer us to the wider world and to the world as a whole.

The central structure of information is a relation of a sign, a thing, and a person: A PERSON is informed by a SIGN about some THING.[1] There are many names for the three parts of this relation. The PERSON can be thought of as the recipient of information, the listener, the reader, the spectator, or the investigator. The SIGN has been called the signal, the symbol, the vehicle, or the messenger. And the about-some-THING is the message, the meaning, the content, the news, the intelligence, or the information (in a narrower and convenient sense I will use frequently).[2] Think of Noah as the recipient, of the dove with the olive leaf as the messenger, and of "the waters were abated from off the earth" as the message. Similarly an archaeologist is the investigator, rocks scattered in a large circle are the signs, and "this is a tipi ring site" is the meaning they convey. You are the reader, the letters are the vehicles, and "this is about information and reality" is the content.

The economy of signs is normally second nature to us and an inconspicuous part of the natural and cultural ecology. It can, however, come sharply into focus in circumstances where an object oscillates between sign and thing or suddenly reverts from reference to presence. A letter is principally a sign. But if it is to matter, it must also be a little thing of some sort, a trace of ink on a page, for example. It is possible to arrange for a playful chatoyancy of a letter between its presence as a thing and its reference as a sign, as in

R R	E	D	U	C C C	E	E
R R	E E	D	U	C	E E E E
R R R R	E E	D	U	C	E E E
R R	E E	D	U	C	E	E
R R	E	D D D D	U	C C C	E	E

The small letters, taken as little things, mere bits of ink, outline the word "holism"; taken as signs and read in sequence, one from each group, they spell out "reduce."[3] A comprehensive view suggests holism while a closer look appears to reduce a thing to its component parts.

```
      O        I
     am        my
    own        way
   of being in
   view and yet
   invisible at
   once Hearing
    everything
    you see I
    see all of
   whatever you
   can have heard
   even inside the
   deep silences of
   black silhouettes
   like these images
   of furry surfaces
   darkly playing cat
   and mouse with your
   doubts about whether
   other minds can ever
   be drawn from hiding
   and made to be heard
   in inferred language
   I can speak only in
   your voice Are you
    done with my shadow
    That thread of dark
    word
     can
     all
     run
     out
     now
      and
       end
        our
        tale
```

This play on signs and things belongs to an ancient tradition of
shaped or figured poems, revived by John Hollander and illustrated
by "Kitty: Black Domestic Shorthair" (fig. 1). Within the silhouette
of Kitty there is the tale of the aloof omniscience of cats and the
inscrutable status of minds, ending with a pun on tale and tail.[4]

Every pun is a play on sign and thing, letting two references
collide in one and the same thing, namely, the sequence of letters
or sounds, with comic or illuminating effect if the pun is good. In
calligraphy, thing and sign are raised to a beautiful balance as they
are on a larger scale in the medieval codex, an illuminated manuscript
on parchment pages, bound between covers ornate with gold and
precious stones.

Distress can make a sign revert to a mere thing. Hikers who were lost in an early fall blizzard on Glacier Park's Chief Mountain had no use for the signifying power of J. Gordon Edward's *A Climber's Guide to Glacier National Park*. They needed paper and tore out a third of the *Guide*'s pages to start a fire. Nor were dollar bills any good as legal tender to skiers who got side-tracked just south of the Park at the Big Mountain ski resort. Instead, the bills served as pieces of paper to light a campfire.[5]

Finally a sign can lose its reference when it is presented at an unnatural scale. As you approach Missoula from the west, you see an *M* clearly outlined on the distant hillside that rises above the University of Montana. If you take the zig-zag path from the campus up the slope to the site of the *M*, you eventually come upon slanted expanses of wavy, white-washed concrete extending in bands 15 to 25 feet wide and up to a hundred feet in length. You will find people picnicking or basking on it. But where is the *M*?

Objects may be too small as well as too large to be evident as signs. Those on a compact disc signifying a text, a piece of music, or a picture are laid down on a spiral track several miles in length and less than a hair wide. The track consists of pits of various lengths and of stretches of unbroken surface also differing in length. Pits represent one or more zeros, the ridges in between stand for ones. But of course the inscriptions are microscopically small, and so the disc that bears them reveals no signs to an unaided human. It is just this round iridescent thing that faintly shows the spiral track.

An object is not a sign or a thing simply; it depends on the context whether it is one or the other. The context is proximally shaped by our playfulness, our needs, our standpoints. Our purposes, however, respond finally to the ultimate context, reality itself, whose cosmic or divine meaning is disclosed by things like landmarks. It is the consonance of cosmic context and focal things that makes the world semantically coherent and allows references to emerge and submerge. Hence information has to be a relation of at least four terms: A PERSON is informed by a SIGN about some THING within a certain CONTEXT.

But even if signs and contexts are plain, one more ingredient is needed if the signs are to tell a person what they are about. Linear B

CHAPTER TWO

is a script found on Crete, dating back to the Mycenaean period—the second millennium B.C.E. But for over fifty years after archeologists had begun to unearth it, no one knew whether the symbols stood for concepts or syllables, or were letters, nor did anyone know what language they conveyed. What was lacking was intelligence in the sense of both factual knowledge and ingenuity. The decipherment became possible when the acumen of Michael Ventris allowed him to extract from the cultural and natural context we share with the writers of Linear B the background knowledge needed to understand the signs of the script.[6] Part of the cultural context is the nature of syllabic writing, the conventions of pictographic writing (as when the shape of a woman or a wheel is reduced to a few lines), and the workings of inflected languages. Part of the natural context is the location of towns and harbors.

Since the script contained eighty-seven distinct signs, the latter could not be logographs, that is, signs such that each stands for an entire thing or concept because reality contains many more than eighty-seven objects or ideas. Logographic systems—Chinese writing is an example—have at least a thousand signs and may have tens of thousands. Alphabetic systems, to the contrary, have always had between twenty and thirty signs. That leaves a syllabic system where each word is decomposed into syllables, each of which is given a distinctive sign. As for the meaning of these signs, some exhibit logographic origin and represent recognizable outlines of a man, a woman, a horse's head, a cup, a spear, or a wheel. But although such resemblances provide a clue to the meaning of the sign, what did the language sound like that was encoded this way?

Scholars had been looking for inflections, those suffixes that distinguish singular from plural, present from past tense, and feminine gender from masculine. Ventris also looked for the names of towns we know to have been celebrated in antiquity—Knossos, site of the royal palace, and Amnisos, the harbor town. Scholars had generally agreed that Linear B could not be a written form of Greek. But as Ventris made headway in his investigations, the improbable became inevitable. Linear B is in fact the syllabic writing of an archaic form of Greek.

It is not true, then, that any person, faced with a sign in a certain context, can gather what the sign is about. It takes intelligence to do so, normal intelligence for customary signs, unusual intelligence when the signs are extraordinary. Hence intelligence with its native and acquired layers is the fifth of the terms that, suitably related and instantiated, produce information: INTELLIGENCE provided, a PERSON is informed by a SIGN about some THING within a certain CONTEXT. There is a pleasing symmetry to this relation. At its center is the sign, the fulcrum of the economy of information, and on it revolves the relation that mirrors the symmetry of humanity and reality, of intelligence and context, that undergirds every kind of epistemology and was first noted by Aristotle in his celebrated formula: "The soul is somehow everything."[7]

The Unsurpassable Eloquence of Reality

In the modern period, this symmetry is thought to have split down the middle. The resulting division of mind and world seems to have given humans dominion over reality. I am here, the world is over there, and at least in principle, it seems, I could construct a complete account of what I have in front of me. Had I but world enough and time, I should be able to write a complete history and draw a complete map of the world. I could begin with the universe at large and gradually work my way to the time and place I happen to occupy. To write the universal history, I put pen to paper and start with the big bang, chronicle the unfolding of the universe and the origin of life, note the evolution of the human race, describe my own birth, and in time conclude my narrative. The account must be summary, of course, to catch up to the present. Many details have to be omitted. But there is one piece of historical reality that will elude my account, no matter how detailed—the event of my writing the account. The problem is the one that besets the project of the complete autobiography. However complete, it will have to omit an account of how it was written. If the autobiographer were to append a postscript, "How I Wrote My Autobiography," the writing of the postscript would elude the chronicle, and so ad infinitum.[8]

As with time, so with space. To draw the complete map of the universe, I might employ the marvels of geographical information systems—devices for sensing, recording, storing, and processing everything in heaven and on earth, all of the information subject to manipulation and display on my personal computer. I begin with the cosmos entire. My computer screen shows the clusters of galaxies. I zoom in on the one that contains the Milky Way, proceed to our solar system, planet Earth, the American continent, the valley I live in, my house, the room I am at work in.[9] I see myself sitting in front of my computer, working with mouse and keyboard, staring at a screen that shows me sitting in front of my computer, working with mouse and keyboard . . . and so ad infinitum. Whether or not one thinks of the infinite sequence of screens showing screens as an irreparable hole in the picture of the universe, there is one region in space that forever escapes representation on the screen—the standpoint from which the picture is seen. The camera that records me in front of my computer must remain invisible. If I use another camera to catch the first, the second recedes into invisibility. The problem here is the one that plagues the universal mirror. It can show me everything in the world, me included. But it cannot show me the universal mirror itself.

Time and space have always and already engaged and surpassed me. The symmetry of humanity and reality is encompassed by reality. Within the world of eloquent and interrelated things and intelligent persons, it is possible to specify the reference of a thing, that is, to let one thing refer to another in a special way and make the former into a sign. Signs are always and already meaningful things. We can discover, explain, and qualify their meanings. But there is no such thing as the original bestowal of meaning on a meaningless sign.[10]

Confusion on this point has sent many a philosopher on a fool's errand.[11] In the culture at large the ties between persons, signs, and things can be severed no more successfully than in philosophy. They can be loosened, however, and lead to a kind of enfeeblement reminiscent of the fate of Antaeus who drew his strength from touching the earth and was invincible as long as the ground renewed his vigor. When Hercules lifted him up and made him lose touch with reality, his fate was sealed. At the same time, when the story of Antaeus was first told in ancient Greece, it was hard to escape the claims of the world. The ground of reality had an engaging force that no one could rise above without a Herculean effort.

Chapter Three

Ancestral Information

Natural Signs and Significance

The ancestral environment of the human family was a world of chiefly natural information. This world is also referred to as the environment of evolutionary adaptation to indicate that our typical genetic endowment has been shaped, tested, and refined in a social and tangible setting that is quite different from ours. What we know of the ancestral environment and its inhabitants comes from the late and final stage of evolution, from the hunting and gathering cultures that give us the earliest complete picture of the human condition. Those cultures have been a distant memory in European history, removed from the present by some ten to fifteen thousand years. But in North America and elsewhere, the stories and traces of the life of hunting and gathering are in some cases separated from our time by little more than a century.

The ancestral environment is the ground state of information and reality. Human beings evolved in it, and so did their ability to read its signs. It is reasonable to assume that the attunement of humans to their original environment felt good. Pleasure is what one feels when things are going right. Information about reality is the way humans are attuned to their wider world. Hence we may look to the ancestral environment to find out about the basic and deeply pleasant structure of information.[1]

One must be careful not to idealize the original and natural setting of information. By our lights, it was limited in its understanding of reality, imperiled by hunger and injury, and ever and again disrupted by the cruelty of warfare. Yet it had for the most part a coherence, order, and vividness that we rarely experience in the contemporary setting of information.[2]

The ancestral environment was profoundly coherent because of the regular interplay of signs and things. When a band of the Salish, some two hundred or two thousand years ago, moved from its summer camp in the Missoula valley north to a winter camp by the "Stream of the Little Bull Trout," now called Rattlesnake Creek, the distant narrows where the creek turns east must have been the first sign they followed. Once they had reached that area, a western tributary to Rattlesnake Creek would alert them that they were within a few hundred feet of the campsite. Finally cairns, tipi rings, or remnants of brush and hide shelters marked the place where they would winter, protected from the punishing east winds and just a few hundred feet below the level to which game retreats from the snow.

Natural signs disclose the more distant environment, yet they do not get in the way of things. A natural sign, having served as a point of reference, turns back into a thing, not as artfully as John Hollander's shaped poems nor as abruptly as the hikers' map or the skiers' money, but naturally and quietly. Thus the ancestral environment, however and wherever humans moved in it, maintained a focal area of presence with a penumbra of signs referring to the wider world.

The ancestral environment of the Salish was well-ordered as well as coherent because some natural signs stood out as landmarks from among the inconspicuous and transitory signs of creeks, rocks, trees, and tracks. Landmarks were focal points of an encompassing order. Snow-capped Lolo Peak was the sign that pointed the Salish toward the salmon on the other side of the Bitterroot divide.[3] The massive portal of Hellgate Canyon marked the passage up the Clark Fork and Blackfoot rivers across the Continental Divide to where the buffalo were hunted.[4] Round and massive McLeod Peak at the north end of the Rattlesnake valley marked the site of vision quests. These landmarks were at once crucial signs and eminent things. They gave order to time and space, distinguishing home ground from hunting and fishing grounds, everyday ground from holy ground, buffalo season from salmon season and bitterroot season. To pass beyond one's landmarks was to lose orientation. In one of the

Montana Salish stories, Coyote, the helper of the Great Spirit, had been traveling west. "His wanderings had taken him far from landmarks he knew," we are told, "and here in the land of sunset everything was strange."[5]

Just as signs and things were naturally and intimately related in the ancestral environment, so were information and knowledge. To recognize a sign was to know what it meant. To appreciate the small cylindrical leaves of the bitterroot flowers as a sign was to realize that they concealed an important thing, the tuber that was a staple of the Salish diet. To notice a young ponderosa pine that had its bark stripped some five feet off the ground was to identify it as a deer rub and to know that here a buck had just cleaned the velvet off its antlers.[6]

So to be intimate with one's world must have been a pleasure when people were healthy, food was plentiful, and conditions were peaceful. And even when the Blackfeet descended from the eastern slopes of the Missoula valley to steal and to kill, the world of a Salish was no less coherent and clear if painfully and bitterly so.[7] Pleasure and pain have not been dissipated by technological information, but they have become confused and distracted through the ubiquitous intrusion of signs into the presence of things. By now we are so inured to the blight of untrammeled information that it takes a deliberate withdrawal to something like the ancestral environment if one is to notice the damage done. Bill McKibben did so by counterposing twenty-four hours of television to twenty-four hours in the natural setting of the Adirondacks.[8] An emblem of the dissolving nearness of things and of the ensuing restlessness that haunts our culture is "buying simply for the pleasure of the act itself," as McKibben puts it.[9] Television advertising constantly abets our belief that ever new bits of property can make up for our failure to appropriate the focal area of our lives. Here the ancestral world offers a salutary contrast. As McKibben explains, "[w]hat sets wilderness apart in the modern day is not that it's dangerous (it's almost certainly safer than any town or road) or that it's solitary (you can, so they say, be alone in a crowded room) or full of exotic animals (there are more at the zoo). It's that five miles out in the woods you can't buy anything. There's no way to drive someplace and spend some money; you can't even phone a TV channel and place an order."[10]

Being out in the wilderness restores one's sanity and serenity. Hikers and backcountry riders tend to be both emphatic and inarticulate about this. Not so Jack Turner who gathered some of the crucial issues into these five sentences: "Alone in the forest, time is less 'dense,' less filled with information; space is very 'close'; smell and hearing and touch reassert themselves. It is keenly sensual. In a true wilderness we are like that much of the time, even in broad daylight. Alert, careful, literally, 'full of care.' Not because of principles or practice, but because of something very old."[11]

A look at the ancestral environment not only shows how a certain order of signs and things makes for a coherent and well-articulated world, it can also reveal that the economy of information depends on a certain character of reality, one we may neutrally call *significant structure*. Philosophers and information theorists have often noted that reality inevitably surpasses and eludes the economy of information. Not that reality is at bottom unstructured and featureless. On the contrary, there is too much structure. Whenever we try to explain or convey something, there is structure, more structure, no end of structure. Thus the task of explanation or information is one of extracting from all available structures the significant structure. In the case of explanation, Carl Hempel puts the problem this way: "Sometimes the subject matter of an explanation, or the *explanandum*, is indicated by a noun, as when we ask for an explanation of the aurora borealis. It is important to realize that this kind of phrasing has a clear meaning only in so far as it can be restated in terms of why-questions. . . . Indeed requests for an explanation of the aurora borealis, of the tides, of solar eclipses in general or of some individual solar eclipse in particular, or of a given influenza epidemic, and the like have a clear meaning only if it is understood what aspects of the phenomenon in question are to be explained."[12] There is a similar problem in music as Diana Raffman has explained. The musical "grammar" that allows a listener to grasp and describe the structure of music is always unequal to all there is in music since "the structural description fails to capture the *non*structural features of the piece. Intuitively speaking, it fails to capture those dimensions of the musical signal which vary along the continuum."[13] As Raffman goes on to show, however, the nonstructural parts of music are not unstructured; it just is impossible to devise a system of digital signs for them.

Speaking most directly about information theory, Fred Dretske contrasts the sharp-edged information of language, which he calls *digital*, with the diffuse information, called *analog*, that a picture conveys—not to mention the boundless information that reality displays. "The cup has coffee in it," gives you just so much information. A picture, to the contrary, "tells you that there is some coffee in the cup by telling you, roughly, how much coffee is in the cup, the shape, size, and color of the cup, and so on."[14] There is as much information as there is structure, too much in other words. "Information, as we usually encounter it," says Keith Devlin, "is not unlike a 'bottomless pit,' seemingly capable of further and further penetration."[15] Thus to get from the abundance of structural information to the economy of significant information, one needs to convert the analog abundance and confusion into digital and significant order. "Digital conversion," Dretske says, "is a process in which irrelevant pieces of information are pruned away and discarded."[16]

In all these cases, it is taken for granted that it is humans who select aspects, pick out structures, and trim away excess to get from mere structure to significant structure. A metaphor, frequently used by Devlin and others to convey this shaping power, has the human subject "carve up the world into entities such as individuals, locations, relations, situations and types."[17] If it is true that reality has fallen silent and declined to the level of structural information, it is reasonable to assume that semantic energy can only flow from subject to object, from humans to reality.

For the Salish and the Native Americans generally, however, meaning flowed in both directions, from subjects to objects and from objects to subjects. So it does and in fact must, even when the subjects are philosophers and the object "is a kapok-like mass of information," to use Devlin's words.[18] Somehow the texture of a thing like the northern lights or of an event like breakfast must come to the philosophers' attention before they can begin to worry about what aspect to select or what information to extract.

To be sure, some meaningful structures are formed by humans. But as a guiding metaphor for the rise of meaning, the picture of material being shaped is misleading. It is more an issue of foreground and background, of emergence and eloquence. And in the ancestral environment particularly, things did not just present themselves minimally and furtively. They were alive with eloquence. "In traditional American Indian cultures," as Baird Callicott has it, "the animals and plants were commonly portrayed as members of a Great Family or Great Society. They were 'persons' worthy of respect, even affection."[19] When in the early 1870s the bison were slaughtered on the southern plains, the animals meant money to some and aroused pity in others. "But for all their differences," Richard White notes, "those who saw animals as commodities and those who saw them as objects of sentiment stood on the same side of a cultural divide. On the other side was a world in which animals were persons and pity was the sentiment that animals felt toward humans. This earlier West appears to us now at once recognizable and utterly strange. Remembering it, we may feel like Dorothy remembering Oz. Because once, when animals were persons, the West was a biological republic."[20] In the earlier West, things stood out from their background so vividly that they appeared to speak to humans. Bears, coyotes, blue jays, and meadowlarks addressed one another and occasionally humans. But rocks too could speak and listen, and an entire valley could show itself to be the remnant of a gigantic rattlesnake as was said of the Jocko valley in western Montana.[21] Such eloquence could rise to the spiritual and culminated in Amortken, "he who sits on top of the mountain," creator of the world and son of the powerful woman Skomeltem.[22]

The eloquence of things makes it possible for signs to be *about-some-thing*. A sign cannot contain a thing entire; but, given human intelligence, it can convey and provoke the impression a thing would leave on a person. In the ancestral environment, the message of a sign is sent by a thing rather than selected by a person, though the recipient needs the capacity to gather the message from the sign.[23]

Conventional Signs and the Fullness
of Information

What landmarks convey can usually be inferred from their context. In less spectacular cases too, immersion in the surrounding reality helps us to grasp the meaning of a sign. Trees that in places had their bark stripped down to the nutritious cambium and have since grown rounded scars around their ancient wounds point to a prehistoric winter camp nearby. Rocks that have been collected into a small circle and contain traces of ash and charred wood under the grass and forbs mark the site of a camp. This intimacy with its context defines a sign as natural or incidental.

A sign becomes conventional and intentional when its message exceeds what can be gleaned from its surroundings. The Medicine Tree west of Missoula, a large ponderosa pine, was the reminder of a young Salish who, pursued by his enemies, was invincible as long as his medicine bundle hung from a limb of the tree and was instantly killed when one of his pursuers succeeded in snatching the bundle away.[24] The tree and its setting were part of the event the tree referred to, but only a part, and the full event needed a narrative convention to be remembered. Thus the tree was more than a merely natural sign and distinguished as a conventional one by the things the Indians hung on it, "small articles of bead work, bear claws, strips of red cloth, queer-shaped stones, bunches of white sage, pieces of buffalo scalp, small pieces of bone, etc."[25]

Once set apart from its natural surroundings, a sign is no longer incidental, something we overhear as it tells its tale; it is now an intentional sign, one that addresses us directly and means to be understood.[26] A cairn is likewise an intentional sign, a pile of rocks that by its density and steepness is set apart from its surroundings and conveys its meaning through a convention without which it means nothing. "Cairns may have been used to mark trails, burials, food caches, or simply the location of an event," say Milo McLeod and Douglas Melton.[27] But in most cases, the convention that was the foundation of a cairn has been lost.[28]

In the ancestral environment of the Old World, the ties between the intelligence of humans, the disclosive power of intentional signs, and the eloquence of things emerge most clearly and eminently in the story of Abraham and Sarah. More important, the story is one of the earliest and most revered documents to raise the question of how encompassing and powerful information can ultimately be. When things begin to speak, they acquire personality. In Genesis, eloquence rises to its highest in divinity and its most personal in the voice of God. Remarkably, the hearer of the voice did not live in anything like the ancestral environment. As Genesis tells it, Abraham had traveled with his father from Ur to Haran, prosperous and literate towns in Mesopotamia at the time when Linear B was being used in Crete, i.e., in the second millennium B.C.E. The Babylonian culture of that period was the most sophisticated in the world, possessing not only a highly developed script, but also a productive hydraulic agriculture, accomplished manufactures, far-flung trade, magnificent architecture, and intricate administrative structures.[29] But the commanding presence of the divine voice suddenly spoke to Abraham and summoned him from the civilization of Babylon to a kind of environment and life that was closest to the ancestral condition of humanity. Without introduction, chapter 12 of Genesis begins:

> Now the Lord had said unto Abram, Get thee out of thy country, and from thy kindred, and from thy father's house, unto a land I will shew thee:
> And I will make of thee a great nation, and I will bless thee, and make thy name great; and thou shalt be a blessing:
> And I will bless them that bless thee, and curse him that curseth thee: and in thee shall all families of the earth be blessed.
> So Abram departed, as the Lord had spoken unto him.[30]

In a passage like this, the People of the Book see the origin and fullness of information. Spirit informs reality. What is needed to grasp this is not just intelligence but faith. Accordingly, Abraham is often called the man of faith. What the faithful receive is not some sign, but the word of God, and that word is not about some thing but regards the beginning and end of all things—divine authority and the fate of humanity. Finally the message is not encompassed by a certain context but by the spirit of God. Nonetheless, what is at issue here is information, not presentation. The nearness of God is audible rather than visible. What Abraham encounters is the word and not the face of God. In Genesis to be sure, there are still places where divinity is embodied in things or persons. When Abraham sat at the door of his tent in the rising heat of the day, God met him in three men or perhaps as a man with two companions.[31] To consummate the covenant, God appeared as a smoking oven and a flaming torch, passing between the slaughtered and divided animals that signified the distinction and union of the contracting parties.[32]

Later passages in the Hebrew Scriptures, however, emphasize that a human being cannot come face-to-face with God. When in Exodus Moses begs to see God's glory, he is told:

> Thou canst not see my face: for there shall no man see me, and live.
>
> And the Lord said, Behold, there is a place by me, and thou shalt stand upon a rock:
>
> And it shall come to pass, while my glory passeth by, that I will put thee in a clift of the rock, and will cover thee with my hand while I pass by:
>
> And I will take away mine hand, and thou shalt see my back parts: but my face shall not be seen.[33]

More emphatic yet is the account in 1 Kings of how God speaks to the disconsolate Elijah:

> Go forth, and stand upon the mount before the Lord.
>
> And, behold, the Lord passed by, and a great and strong wind rent the mountains, and brake in pieces the rocks before the Lord; but the Lord was not in the wind: and after the wind an earthquake; but the Lord was not in the earthquake:
>
> And after the earthquake a fire; but the Lord was not in the fire: and after the fire a still small voice.
>
> And it was so, when Elijah heard it, that he wrapped his face in his mantle, and went out, and stood in the entering in of the cave. And, behold, there came a voice unto him, and said, What doest thou here, Elijah?[34]

That there should be information of such magnitude seems incredible or incomprehensible to some of the most thoughtful people today. When people of scientific temperament contemplate reality, they are chiefly impressed by the general lawfulness of its structure. Moreover, some of the most prominent scientists regard it as possible and desirable to reach the fullness and completion of our understanding of structure and lawfulness through a final theory, one that would unify all forces of nature.

That the significance and eloquence of reality should reveal itself with similar finality seems impossible to grasp. It is as though Abraham received the big bang of divine information, while we are left with faint cosmic background radiation. The meaning of reality has declined and become occluded. It has been reduced to contingency—the unexplainable residue of accident and randomness. That divinity, whether from within or beyond the world, should reveal itself in contingency seems inconceivable, and so does the idea that lawfulness for its part—for example, in its mathematical purity— should be a trace of divinity, a trace that seemed indubitably divine to Augustine (354–430) and Leibniz (1646–1716).

Thus Steven Weinberg, who has thought deeply about the physical lawfulness of reality and has made significant contributions to its discovery, has said, "It would be wonderful to find in the laws of nature a plan prepared by a concerned creator in which humans played a special role. I find sadness in doubting that we will."[35] Similarly Daniel Dennett, who has given a spirited account of the lawfulness of evolutionary theory, has written, "I certainly grant the existence of the phenomenon of faith; what I want to see is a reasoned ground for taking faith seriously as a *way of getting to the truth,* and not, say, just as a way people comfort themselves and each other (a worthy function I do take seriously)."[36]

And yet both Weinberg and Dennett attest to faint echoes of divinity, to the gratuitous beauty and sacred magnificence of reality. Weinberg has said, "I have to admit that sometimes nature seems more beautiful than strictly necessary. Outside the window of my home office there is a hackberry tree, visited frequently by a convocation of politic birds: blue jays, yellow-throated vireos, and loveliest of all, an occasional red cardinal. Although I understand pretty well how brightly colored feathers evolved out of a competition for mates, it is almost irresistible to imagine that all this beauty was somehow laid on for our benefit."[37] And Dennett has written, "Is something sacred? Yes, say I with Nietzsche. I could not pray to it, but I can stand in affirmation of its magnificence. This world is sacred."[38]

In any event, people anciently have found it necessary to realize information, whether divine, cosmic, or mundane, in some tangible and permanent way. As Victor Hugo has it:

> When the memory of the first races felt itself overloaded, when the mass of reminiscences of the human race became so heavy and so confused that speech naked and flying, ran the risk of losing them on the way, men transcribed them on the soil in a manner which was at once the most visible, most durable, and most natural. They sealed each tradition beneath a monument.
>
> The first monuments were simple masses of rock, "which the iron had not touched," as Moses says.[39]

Thus when Abraham first reached Canaan and God had repeated the promise and shown him the land, Abraham's response was to build there "an altar unto the Lord, who appeared unto him."[40] Abraham did so again some twenty miles south, on a mountain east of Bethel, and once more, after his return from Egypt, by the oaks of Mamre, north of Hebron.[41] And so, in fact, did Isaac and Jacob when God appeared to them.[42]

Similarly, the ancestors or predecessors of the Salish in western Montana had been marking places and events at the time Abraham built altars and thousands of years before.[43] The stone monuments of the Blackfeet and their predecessors on the northern plains are particularly expansive and elaborate. These "medicine wheels" were centered on a cairn, up to twenty feet in height, and had spokes and rings extending a hundred feet or more.[44] Their particular functions too remain conjectural. But we know enough from the biblical and Native American traditions to suggest that the lapidary signs of altars, cairns, and circles amplified and stabilized the eloquence of divinity and reality. They marked a center of celebration, the origin of a covenant, the course of a trail, the location of a camp, the grave of an ancestor, or a place where peace was made. Thus, we may conjecture, monumental signs contributed to an environment that at its normative best constituted a coherent, well-ordered, and eloquent world.

Such a world comes briefly to light in Genesis when Abraham, having parted ways with his nephew Lot, moves to the eastern part of Palestine. Abraham's world is rendered from the outside in. First, conveyed by the word of the Lord, comes the grand context of space and time, the expanse of the land and the future:

> And the Lord said unto Abram, after that Lot was separated from him, Lift up now thine eyes, and look from the place where thou art northward, and southward, and eastward, and westward:
> For all the land which thou seest, to thee will I give it, and to thy seed for ever.
> And I will make thy seed as the dust of the earth: so that if a man can number the dust of the earth, then shall thy seed also be numbered.
> Arise, walk through the land in the length of it and in the breadth of it; for I will give it unto thee.

Within this context, the focal area of Abraham and Sarah's life is designated by their tent and symbolized by an altar: "Then Abram removed his tent, and came and dwelt in the plain of Mamre, which is in Hebron, and built there an altar unto the Lord."[45]

Another instance of the ancestral world opens James Welch's *Fools Crow*. The account, albeit more artful, is similar to that in Genesis. It begins by tracing the context of time, signaled by the signs of the weather: "Now that the weather had changed, the moon of the falling leaves turned white in the blackening sky and White Man's Dog was restless. He chewed the stick of dry meat and watched Cold Maker gather his forces. The black clouds moved in the north in circles, their dance a slow deliberate fury."

Next we are shown the space of nearness, marked out by the tipis of a band of Blackfeet: "It was almost night, and he looked back down into the flats along the Two Medicine River. The lodges of the Lone Eaters were illuminated by cooking fires within. It was that time of evening when even the dogs rest and the horses graze undisturbed along the grassy banks."

The camp is then set in the wider context that is oriented along the Continental Divide and dominated by a mountain: "White Man's Dog raised his eyes to the west and followed the Backbone of the World from south to north until he could pick out Chief Mountain. It stood a little apart from the other mountains, not as tall as some but strong, its square granite face a landmark to all who passed."

FIGURE 2 Karl Bodmer, *Assiniboin Medicine Sign* (1833)

Chief Mountain is topped by a vision quest site, defined by rocks and buffalo skulls, reminiscent of the cairn Karl Bodmer had painted in 1833 (fig. 2), some thirty years before White Man's Dog was contemplating his world. This sacred monument on Chief Mountain, finally, opens up a realm of past heroes and future possibilities, challenges that at the time seemed forbidding to the novel's protagonist who later gains fame and is renamed Fools Crow:

> But it was more than a landmark to the Pikunis, Kainahs and Siksikas, the three tribes of the Blackfeet, for it was on top of Chief Mountain that the blackhorn skull pillows of the great warriors still lay. On those skulls Eagle Head and Iron Breast had dreamed their visions in the long-ago, and the animal helpers had made them strong in spirit and fortunate in war.
>
> Not so lucky was White Man's Dog. He had little to show for his eighteen winters.[46]

For all their orienting power, rock monuments were hard signs. Being ponderous and place-bound, they could not be used to convey information from place to place, and though they were massive and could be monumental, their capacity to convey information was slight. To the uninitiated they say little more than, "Yes, there is a message here," while the bare or natural surroundings seem to say, "No, there is no message elsewhere." Strictly as a sign, a cairn and its surroundings constitute just one bit of information. A cairn is not nearly as articulate and explicit as the petroglyphs and pictographs of the northern Rockies.[47]

The something a cairn conveys arises from a particular occasion, most eminently the appearance of the divine. Since the informational capacity of the cairn is so small, large tasks remain for the people to whom the cairn is significant. Elaborate instruction and careful memorizing are needed if the message of the sign is to survive. Moreover, since the cairn as a sign is occasion-bound as well as place-bound, it normally carries one and only one message. It is not a vessel that can be used to convey different contents on different occasions.

All this suggests why conventional signs developed from the stationary and monumental to the mobile and instrumental and so became better signs if lesser things. In fact the slight and footloose signs we call letters turned out to be much more reliable and durable containers of information than monumental signs of stone or bronze. Had not, a few centuries after the Abrahamic events, a writer now commonly called J set down for us the account known as Genesis, God knows whether Abraham and Sarah's story would have come down to us.[48] But evidently the instrumental signs in turn depend on the monumental. The latter make for information that is worth recording and transmitting, and even the mundane information that occupies much of daily life today moves along lines that have been laid down by focal events and monumental signs.

Chapter Four

From Landmarks to Letters

Remembering before Writing

If we think of information as a relation—INTELLIGENCE provided, a PERSON is informed by a SIGN about some THING in a certain CONTEXT—we can hardly fail to notice that in a hypertrophically informed society like ours the SIGN looms large. If all our ledgers, accounts, files, calendars, memos, letters, scores, plans, maps, data bases, and other records should suddenly disappear, so would the order and coherence of our society. We are so used to the mass and sophistication of our vehicles and containers of information that a society without them seems primitive and incomplete. We used to call such a culture illiterate. Within the twentieth century, however, we have learned to recognize that life without letters has its own coherence and dignity, and we have come to call it oral rather than illiterate.[1]

Just as the notion of information sheds revealing light on cultures where information, though present, was unnoticed as an explicit phenomenon, far less as a cultural problem, so writing, by its absence, makes oral cultures specially remarkable and, at least for us, raises the question as to what lent a preliterate society the stability and continuity we owe to written records. The general answer is that in an oral culture the relative weakness of signs was balanced by more robust intelligence, a fuller engagement of the person, and greater intimacy of the context.

Intelligence in this instance is the capacity to retain information. To lend coherence to the texture of its works and days, an oral culture had to rely on memory—a small and unsteady loom. To enlarge and to brace it, oral people supported it with shared recollections, fastened it to bodily rhythms, and above all surrounded it with intentional and conventional signs that bespoke a real context.

Memorable events were anchored in several memories at once. The story of how Abraham bought a burial place from Ephron recounts a careful oral agreement witnessed by many. "And the field of Ephron, which was in Machpelah, which was before Mamre, the field, and the cave which was therein, and all the trees that were in the field, that were in all the borders round about, were made sure unto Abraham for a possession in the presence of the children of Heth, before all that went in at the gate of his city."[2] To record, at first, meant to bear witness, and to recall a crucial date or settle a dispute about an ancient privilege, a predocumentary community could resort to the memory of respectable witnesses, solemnly interrogated.[3]

Minstrels and remembrancers, who as a matter of profession had to memorize and recall expansive pieces of language, engaged the movements and rhythms of the body to hold on to songs, epics, histories, and instructions.[4] To the motions of the body and the meter of language correspond patterns of content and meaning, formulas that capture a person, a thing, or a setting in a memorable phrase that is used over and over again. The Homeric epics, committed to writing at the very beginning of alphabetization, convey to this day the stately, recurring, and reassuring formulas of the noble, much enduring Odysseus; of the goddess of the flashing eyes, Athena; of the wine-dark sea; of meat galore and mellow wine.[5]

Memory, finally, was joined to external markers. There are of course the early lapidary signs, the altars of the biblical patriarchs, the ancient Greek stone markers that defined boundaries, indicated mortgages, or called attention to tombs and temples, and the cairns of the Native Americans on the plains and in the mountains.[6] But all kinds of things served as reminders. To memorialize the conveyance of a piece of property in the early Middle Ages, "a symbolic object, such as a knife or a turf from the land" was transferred.[7] Similarly, in the words of a medieval lawyer, if "livery is to be made of a house by itself, or of a messuage for an estate, it ought to be made by the door and its hasp or ring, by which is understood that the donee possess the whole to its boundaries."[8] In an early and touching piece of anachronism, if the story is to be trusted, the Earl Warenne, being unable to offer Edward I's judges a written warrant of his estate, produced a sword from the Norman conquest and said, "This is my warrant."[9] A cup, a gold and ruby ring, a staff cut "from the land" similarly served as markers and warrants, and so do wedding rings to this day.[10]

Counting

But even in the most ancient cultures, signs had been devised to lighten the burden on memory and loosen the ties to context. Those signs were counting devices, and the earliest that have been left us are tallies, bones with parallel incisions that were used to record phases of the moon, perhaps, or animal kills. The oldest dates from about 100,000 B.C.E.[11] The practice of tallying was alive down to the Middle Ages where hazel-wood sticks rather than bones were used.[12] In some medieval instances, the location of the tally specified what was counted. In a stable it was the yield of milk; in a mill, bags of corn; in an exchequer's office, payments into the treasury. General conventions determined the how-much of payments; large notches stood for pounds, medium ones for shillings, and small ones for pennies. From the late thirteenth century on, writing on the tally sticks was used to spell out what previously was left to circumstance and agreement.[13]

There must have been other devices in prehistoric times to keep track of discrete items—sets of pebbles, twigs, grains, beans, or shells.[14] Mesolithic pebbles, marked and colored with iron peroxide, have been found in Southern France, and some Iraqi shepherds reportedly keep track of their animals by means of clay pellets to this day.[15]

Counting devices disburdened human intelligence from having to remember the number of whatever, and they liberated human intelligence from the confinement of contexts because, unlike landmarks and monuments, tallies and counters could be easily moved from place to place. But the price of disburdenment and mobility was a diminishment of information. A counting device was not the vehicle and reminder of a sacred story but merely the indicator of however many kills, sheep, or bags of corn. Writing, and alphabetic writing particularly, appears to combine the best of two worlds. Like counting devices, it liberates humans from the labor of memorizing and from the fetters of circumstance; and like landmarks and monuments, it can convey powerful information.

Counting, it turns out, is the bridge between the monumental signs of cairns and altars and the instrumental signs of letters. The link from piety to literacy was numeracy. Humans began to count before they started to write, and they recorded numbers or perhaps numbered things before they wrote down words or ideas. The fragments of ancient times that have come down to us suggest that counting, the beginning of numeracy, evolved out of the creation myth.

In the stories and reenactments of creation, things came to be as they were called forth by the creator or priest.[16] Often things were called in twos as in Genesis. God brought humans into being when "male and female created he them."[17] Similarly Noah saved the animal realm when "there went in two and two unto Noah into the ark, the male and the female, as God had commanded Noah."[18] Here, perhaps, lies the origin of counting in groups of two, a system that appears to be the oldest and anciently the most widely practiced.[19] Calling comes to counting when the same words are used to call up in turn the members of each pair: one and two.[20] Counting comes into its own when it is extended beyond the number two and greater numbers are expressed as groups of twos and ones, for example, three as "two one," four as "two two," five as "two two one," etc.[21] Ritual counting was sometimes performed by gathering stones to take account of the number of persons of an assembly, or an army; or to mark a place as holy, as the site of a covenant—a covenant with God or a covenant between humans as was the case when Laban settled his differences with Jacob and said,

> Now therefore come thou, let us make a covenant, I and thou; and
> let it be for a witness between me and thee.
> And Jacob took a stone, and set it up for a pillar.
> And Jacob said unto his brethren, Gather stones; and they took
> stones, and made an heap.[22]

The Salish practiced a ritual counting in certain places where travelers were required to pay their respects to some spirit of the land by counting themselves in. A collector of Salish stories tells us: "At some of the passes in the mountains, near a land mystery, each traveler put down a stone; if he did not, he would have ill luck."[23]

For mundane purposes, there was in all cultures some method of keeping track of quantities by means of tallies, by matching one set of things with another, or by simply assigning different words to different quantities of things. Writing, in comparison, is an improbable cultural device and was invented only three times in human history, first in the late fourth millennium B.C.E. of Mesopotamia—whence it spread to Egypt and the Indus Valley—and later in China and in Mesoamerica.[24] In the latter two cases, we do not know in any detail how people came to write. In the first instance, however, we can trace a line from counting to writing and to the uniquely efficient kind of writing by means of letters.

Shepherds in the ancient Near East, keeping count of their charges by means of clay pellets, had a device that did just one thing—keep track of *how many*. Playfulness, perhaps, came to the aid of usefulness when the handling and forming of clay led to regular and distinctive little tokens, spheres, discs, cones, tetrahedrons, ovoids, etc.[25] The first ones appeared around 8,000 B.C.E. in the Near East.[26] For three and a half millennia these plain tokens were used in an agricultural economy to keep track of farm products. An ovoid represented a jar of oil; a cone, a small measure of grain; a cylinder, an animal; etc.[27] Around the middle of the fifth millennium B.C.E., tokens of more intricate shapes, bearing linear and punched markings, began to appear. They were used to represent products of urban manufacture like textiles, metals, jewelry, and perfume.[28]

Clay tokens were more informative signs than mere pellets because their distinctive geometrical shapes and their markings indicated not only *how many* but also *what* and in some cases *how much*. A sphere, for example, stood not only for "one measure [the number] of grain [the content]" but for "one *large* [quantity] measure of grain." In becoming more informative, these signs became more concrete as well and more like things presenting manifold information at one and the same time.

Tokens, moreover, reflect the three-dimensional boundedness of things and the ways they can be collected or segregated. Counting with tokens required therefore that a specific what (and sometimes a specific how-much) was inextricably counted along with the how-many. This has been aptly called concrete counting.[29] There are traces of this in today's English. Twins are not just any two things, but two siblings, conceived at the same time by the same mother. A trio is not just any set of three, but a group of three musicians.

Writing

Because tokens can get scattered and lost almost as easily as the things they are supposed to keep track of, tokens belonging together were enclosed in a clay envelope. But since the container concealed its contents, accountants impressed on the soft clay of the envelope the shapes of the tokens that were enclosed. It was soon realized that the markings on the outside made the tokens on the inside superfluous. In due course, impressions on clay tablets began to replace clay tokens.[30] Thus the three-dimensional reality of the signs was reduced to essentially two dimensions. At the same time the collection of however many signs into a set was no longer merely implied by how and where they were gathered and stored, but was explicitly and definitively indicated by the way the signs were impressed together on the tablet.

Seeing five disks with crosses inscribed on a tablet, some accountant must have realized that one crossed disk and five quick marks would convey the intended information as well or even better.[31] Thus numbers came to be separated from commodities, the former eventually being impressed with the blunt end of a stylus, the latter incised with the sharp end. Abstract counting, the determination of the mere how-many, was born or reborn or, at any rate, for the first time explicitly rendered apart from and next to what was being counted. Later still, not only the how-many (one) and the what (oil) but also the how-much (*sila* or measure) was segregated and stated explicitly.[32] The crucial move in this development was making reference, formerly implied in counting, explicit and distinct. On a clay tablet a circle with a cross stood simply for *sheep*.[33] This slight, abstract, inscribed sign was about some thing. Writing was born.

The sign in this case was a logograph, a symbol that stands for an entire thing, concept, or word. Logographic writing has been the rule in the invention and development of writing. By 3,000 B.C.E. there existed at least half a dozen such systems in the Near East.[34] They work perfectly and admirably, as Chinese writing shows, yet they impose a heavy burden on memory. Like a cairn or a tally bone, each logographic symbol requires for its meaning a convention of its own though, unlike cairns and tallies, the logographic conventions are not bound to particular places and circumstances, but are established once and for all. Still, since logographic systems have thousands of symbols, thousands of separate conventions must be memorized.[35]

Once more playful irreverence must have been the channel of utility. Phonetic writing, depicting spoken sounds rather than intended things, events, or concepts appears to have come about through rebus writing, the kind of script you used to find on the inside of beer bottle caps. In a Heineken commercial, claiming that "Heineken refreshes the parts other beers cannot reach," "other" is represented by an udder and "cannot" by a watering can and a knot in a rope.[36] Rebus writing, however unwittingly or serendipitously, made a bold move of abstraction. To an orthodox reader, a logograph conveyed a thing and a sound at once and indissolubly. The witty and venturesome rebus scribe dissociated the sound from the thing and used the sound as the vehicle for an entirely different thing.

The first employment of rebus writing served to solve this or that particular problem, for example, the rendering of a proper name by way of a logograph that sounded like the name of a person or a god.[37] But real gains of efficiency were not made until words were decomposed into syllables and the rebus principle was brought to bear on them. To use an English example, if you have symbols for SLIP and PEAR you can dispense with the symbol for SLIPPER. Parsimony requires in addition that the vowels be dropped so that SLIP, SLEEP, SLAP, SLOP, and SLOOP are all rendered as SLP. The context determines which of the five is the intended word. If English writing were logographic and had separate signs for BAT, BAIT, BET, BIT, BITE, BOT, BOOT, BOUT, and BUT, a shift to the syllable "BT" would reduce the signs in this instance by a factor of nine. In general, however, syllabic writing leaves a script with hundreds of distinct syllables and signs.

The response to this residual inefficiency was the further decomposition of syllables into letters. It occurred just before the middle of the second millennium B.C.E. somewhere in Syria or Palestine.[38] The details of the process are lost to us. Perhaps it was the prominence at the beginning of words that allowed the letter to emerge as a basic constituent of writing.[39] Certainly the convergence of many forms of writing in this region and at this time and the need for efficient communication to accommodate trade spurred the invention of alphabetic writing. But then again, the spur may have been the desire to provide a powerful tool for the commemoration of Abraham and Sarah's story.[40]

At any rate, toward the end of the second millennium B.C.E., we find inscriptions in a North Semitic alphabet of 22 consonants.[41] The Greeks added vowels early in the first millennium B.C.E. and bequeathed their alphabet to the Romans, who in turn passed it on to the Middle Ages whence it came to modern civilization.[42]

Looking back over the course that the development from landmarks to letters has taken, we might think that it has led us from signs that referred to things in a roundabout way to signs that refer us to signs—letters referring to words. And it appears that writing owes its informative power simply to getting a ride on the powerful information system of oral language. Alphabetic writing seems to be the simple transposition of language from an audible to a visible medium—nothing could be simpler one should think. But the fact that it took a long evolution to get from logographic to phonetic writing and that this path was traveled only once in human history indicates that something more difficult and remarkable was accomplished in alphabetic writing.

To be sure, spoken language is a most intricate system of communication, exhibiting manifold structures and elements, among them those that are captured in alphabetic writing—the components we call sounds or letters, the compounds we call words, and the compounds of these compounds, the sentences. But in speech, these structural features are almost inextricably woven into a rich bodily and contextual reality. Spoken language is not so much a thing that a person uses as it is a representation of the way a person is. Speech is to the mind as skin is to the body. It is the way a person comes to be a definite and expressive creature. Speech is as inseparable (even though distinguishable) from a person's thoughts and feelings as skin is from bones and muscles. And just as bodily movements are fluid, passing, and largely instinctive, so is spoken language.

Moving and speaking persons, moreover, are who and what they are in a specific context. Much of the meaning of posture and gesture is clear from its setting. So it is with speaking. Words like *you*, *now*, *here* particularly mean everything and nothing without a context.

Keeping the richness of spoken language in mind, we can see that alphabetic writing, however serendipitously arrived at, constitutes a radical abstraction from speaking, no less so than an X-ray picture abstracts from a living person. Writing sets aside the fluidity, inflection, evanescence, the embodiment and context of speaking and leaves us with a rigid, permanent, and detached piece of information. In fact, writing extricates information from persons and contexts and sets it off against humanity and reality.

Chapter Five

The Rise of Literacy

Letters vs. Memory

Literacy is the ability to read and write. In a more dramatic sense it is the force that invades and transforms an oral culture. The drama of this encounter has been enacted in ancient Greece, in the early Middle Ages, and in the conquest of this continent by Europeans, among countless other instances. Writing poses a challenge that no preliterate society has been able to evade. Whatever else may be controversial about literacy, its overwhelming power is not. Primary orality has never been able to resist it in the long run.

Writing constitutes a uniquely compact and potent kind of information. Of the five terms of natural information, where INTELLIGENCE provided, a PERSON is informed by a SIGN about some THING within a certain CONTEXT, two all but detach themselves and make the remaining three drift into the background. Writing consists of signs that are about some thing, letters that convey meaning. By itself, writing is not bound to a particular person or context, and its possession requires no particular intelligence.

So, at any rate, Plato thought. As fate would have it, this great thinker witnessed the first flowering of social literacy where knowledge of letters was not just the skill and privilege of a guild of scribes or a group of priests, but rather the possession of an entire class of readers and writers.[1] Plato experienced the rise of such a class and contemplated, with slight jaundice and inconsistency as it turns out, the troubling consequences of this development. In the dialogue *Phaedrus*, Plato recounts the story of the invention of writing and attributes it to the Egyptian god Theuth. Writing, when encountered fully fledged, appears to possess such superhuman power and precision as can only be traced to a divine origin—a view that in one version or another was maintained until the eighteenth century.[2] In this spirit, Theuth takes a bright view of his invention, predicting that it "will make the Egyptians wiser and will improve their memories; for it is an elixir of memory and wisdom that I have discovered."

As Plato has it, however, the Egyptian king and god Thamus, to whom Theuth had made his presentation, was doubtful, and he complained that "this invention will produce forgetfulness in the minds of those who learn to use it, because they will not practice their memory. Their trust in writing, produced by external characters which are no part of themselves, will discourage the use of their own memory within them. You have invented an elixir not of memory, but of reminding; and you offer your pupils the appearance of wisdom, not true wisdom, for they will read many things without instruction and will therefore seem to know many things, when they are for the most part ignorant and hard to get along with, since they are not wise, but only appear wise."[3]

The note of disdain in Thamus's reply is a reflection of the doubt and even scorn Athenians cast on the very idea that something external, something one could take up and lay down at will, would make one a more accomplished person. Young men with books were regarded as ridiculous figures in Athenian comedy.[4] Phaedrus in Plato's dialogue very nearly is one of them. He is both enamored with a copy of the speech he carries around and ashamed of displaying it when he runs into Socrates. More generally, Plato held that writing is not for serious matters.[5] Even laws, once committed to writing, suffer a loss of weight and seriousness.[6] Compared with genuine government, legislation is merely a prudent game.[7]

Yet Plato could not entirely convince himself that the power of literacy was adequately met with ridicule. He recognized that writing compacts the large and living structure of natural information and feared that detached parcels of written information, easily acquired, would take the place of genuine wisdom, arduously earned. Writing, he thought, would promote both vanity and stupidity.[8]

Whether writing, due to its detachment from context and intelligence, actually promoted impertinence the way Porsches do today is

CHAPTER FIVE

questionable. To be sure, writing injects an ambiguity into the notion of *having information*. In the economy of natural information, to recognize signs for what they are is to know what they mean. To have information is to have (indirect) knowledge. But once writing appeared on the cultural scene, it was possible to have information, in the sense of owning a book, without having knowledge, that is, without having read it. Plato thought the ambiguity was insidious in bestowing the prestige of knowledge on people who owned information without having absorbed it. But just as we cannot really get ourselves to believe that a Porsche will transform a dull and paunchy man into a sophisticated handsome devil, so the Greeks soon came to understand that ownership of a book will enhance one's reputation only if one has read the book comprehendingly.

In another respect, however, writing can unequivocally upset the ancestral ways of knowing. In the natural economy of information, signs and things keep a fine balance. Natural signs emerge, refer, and disappear. They do not get in the way of things. Writing, to the contrary, allows for an endless accumulation of information, and unchecked accumulation leads from perspicuity—the signal benefit of natural information—to confusion. This too was an experience had in the very early stages of literacy.

By the time of Plato's youth, late in the fifth century B.C.E., it had been the custom of Athenian democracy to publish its laws and decrees so that "anyone who wants may see them."[9] They were inscribed on walls, slabs, pillars, tablets, and papyrus rolls. The rolls were kept by officials or in the council house. The proliferation and dispersion of laws and records made it all but impossible to know what was in force and what was not.[10] But even after a city archive had been established, it was easier to use the information contained there for partisan purposes than for circumspect reference.[11]

Some 1700 years later, writing and record keeping made its way into the municipal administration of Freiburg in the Upper Rhine Valley. For most of the thirteenth century, the town employed scribes as the occasion required it. Late in the century a full-time scribe took office. At the turn of the century an assistant was added.[12] But, scattered as they were in various rooms of city hall, in one of the towers of the great Gothic church, and in the attic of the House of Merchants, the records that accumulated were of limited use[13] (though several attempts were made to collect the most important material between the covers of one book[14]).

As clumsy as written records were en masse, in the end a particular document on a particular occasion possessed overwhelming power when it confronted oral traditions and agreements. Documents gained this power gradually and irresistibly. They did so in classical Athens.[15] And they did so again after the Norman Conquest of England in 1066. Latin and Anglo-Saxon writing had served sacred and historical ends. But reading and writing were not widely shared and broadly influential practices.[16] The Norman conquerors introduced writing as an instrument of domination and administration. The Domesday Book is a monument to William the Conqueror's attempt at taking firm and complete possession of his conquest though the practical utility of his imposing instrument turned out to be disappointing.[17]

But otherwise writing was developed into a powerful administrative tool. By the twelfth century, the Royal Chancery composed 4,500 letters a year. In the thirteenth century, eight million charters were issued just to fix in writing the hitherto oral understandings that had secured the tenure of smallholders and serfs on their land.[18] By 1300 literacy had reached every village.[19] The year 1189 was officially designated as the divide between memory and written record. Claims from the far side of the watershed could appeal to recollections and traditions; any claim since 1189 had to be based on a written document.[20]

The clash between the oral and literate cultures reached a fine point of controversy when emissaries of Anselm, bishop of Canterbury, and representatives of Henry I returned with different messages from Pope Paschal on who had the right to appoint bishops and priests and on whether Henry had seriously violated the papal rule. Anselm's monks returned with a document in their favor. The king's trusted bishops conveyed an oral message to support their sovereign. Who was to be believed? The monks pleaded the superiority of literacy and extolled the "documents signed with the pope's seal (scriptis sigillo pape signatis)" above "the uncertainty of mere words." The royal party, conversely, set greater store by the testimony of bishops than by "the skin of wethers blackened with ink and weighted with a little lump of lead."[21] Yet medieval scorn no more succeeded in stripping signs of their power of reference and in reducing them to dumb things than had ancient ridicule. Medieval contempt of writing turned to apprehension and sought comfort in Paul's warning to the Corinthians that "the letter killeth, but the spirit giveth life."[22]

Literacy and Community

Writing has a precision and permanence that human recollection, however widely shared, cannot match. When literacy invades an oral culture, it drains vitality from the community. Its members no longer seem to be the ultimate warrantors of validity and order. Writing, in the words of M. T. Clanchy, appears "to kill living eloquence and trust and substitute for them a mummified semblance in the form of a piece of parchment."[23]

The power of letters, of course, cannot literally detach itself from humanity. It needs to be wielded by some person or other. Ignorance of how the power of letters is exercised in a particular culture can engender both unrealistic expectations and uncomprehending despair. The former were entertained by the Blackfoot chief Heavy Runner. In James Welch's telling he made this request of General Alfred H. Sully on January 1, 1870: "[A]nd so I ask the great seizer chief for a piece of paper with writing on it that states that I, Heavy Runner, am a friend to all whites. If your people decide to make war on the Pikunis, I would desire it to be known to them, on paper, that Heavy Runner and his followers are at peace."[24] But when a few weeks later the U.S. Army under the command of Major Baker mistook Heavy Runner's camp for that of the Many Chiefs whom they had set out to punish, letters on paper were to no avail. Welch has a witness to the massacre report, "Curlew Woman says Heavy Runner was among the first to fall. He had a piece of paper that was signed by a seizer chief. It said that he and his people were friends to the Napikwans. But they shot him many times."[25] A. B. Guthrie too recorded the incident to conclude his story of how the nature of the West and the culture of the Indians were driven to submission: "Heavy Runner was trotting out there, waving his friendship paper.

He jerked to a halt and turned and fell as a shot sounded. His paper fluttered away."[26]

The unintelligibly forcible side of the use of letters was encountered by the Native Americans who negotiated treaties with the U.S. government. In July of 1855, Governor Stevens of Washington Territory met with the Salish, Pend d'Oreilles, and Kootenai of what is now western Montana to settle on the location and conditions of a reservation. For the most part the apprehensions of the Indians about the alien and compelling character of a written treaty was overwhelmed by their hope of relief from poverty and from the attacks of the neighboring Blackfeet and by Stevens's repeated insistence that treaties were common and inevitable. Only Big Canoe of the Pend d'Oreille clearly saw the clash of cultures. In a long speech, ironically disfigured by poor translation or recording, he said, "When I lay [lie] down my heart is sad[;] now my chief you say now I am blind if I want to talk. Here are my eyes, my heart, my brain, I study. You white man; there are your eyes lying all over the table, that is the reason you are smart, you always look at your papers; now you talk, it is right when you talk straight. I from my heart and my brain speak."[27]

Plato too had felt the foreign, if not the forcible, impact of letters on the spirit of conversation. He was vexed by the attempts of rivals and students to fix on paper what exercised his mind, and he emphatically denounced writing to uphold the living presence of thought: "There is not now any written paper of mine about it, nor will there ever be one. For it cannot be at all set down like other learning. Rather from intense communion and from living together with this thing it suddenly arises in the soul like a light set off by a leaping fire; and once engendered it nourishes itself."[28] While Plato surely was wrong in fearing that writing would undermine indirect knowledge, that is, the comprehension via signs of matters that are distant in space or time, he rightly saw that detached information can be the adversary of direct knowledge—the acknowledgment of thoughtful and eloquent persons in their immediate presence. Philosophy, as Plato saw it, comes truly alive only when living persons are engaged in an actual dialogue.

If writing can be a tool of alienation and domination, it can become an instrument of liberation as well. The African-American children on Port Royal Island off the coast of South Carolina who had just been freed from slavery sensed the power of literacy deeply and testified to their trust in its liberating power on the occasion of a funeral in the summer of 1862. The event was memorialized by William C. Gannett, one of the "abolitionist missionaries of freedom and Yankee culture" who tried to help the former slaves gain independence and self-sufficiency:

> 'Twas at set of sun; a tattered troop
> Of children circled a little grave,
> Chanting an anthem rich in its peace
> As ever pealed in cathedral-nave,—
>
> The A, B, C, that the lips below
> Had learnt with them in the school to shout.
> Over and over they sung it slow,
> Crooning a mystic meaning out.
>
> A, B, C, D, E, F, G,—
> Down solemn alphabets they swept:
> The oaks leaned close, the moss swung low,—
> What strange new sound among them crept?
>
> The holiest hymn that the children knew!
> 'Twas dreams come real, and heaven come near;
> 'Twas light, and liberty, and joy,
> And "white-folks' sense,"—and God right here![29]

Literacy can be liberating because the detached information of writing is more widely and easily available than natural or oral information. Letters can raise the power of original information and render the world more open and equitable. Writing, however, not only has the capacity of natural information in a heightened version, it has in addition a very nearly novel power. It is a power that Plato, fearful for the welfare of knowledge, was unable to acknowledge. In his effort to secure the dignity of knowledge, he sought to restrict letters to the status of reminders, purely mnemonic signs conveying merely retrospective contents. Only a fool would think, he said, "that written words are of any use except to remind him who knows the matter about which they are written."[30]

But writing not only brings near what is distant in time and space but also allows us to realize what otherwise is prohibitively remote as a possibility of conception and imagination. In detaching information and setting it before us in a compact and enduring form, writing makes information into the material that Plato himself was so masterful in forming. His writings are more than reminders. They are works of art that have powerfully defined the compass of occidental philosophy. So powerful has his instruction been that, as Whitehead famously said, "the safest general characterization of the European tradition is that it consists of a series of footnotes to Plato." [31]

Truly oral speaking is like gathering stones. You take what is at hand and comes to mind. If things go well, if the pieces are colorful, polished, and fit together, speech can rise to monumental proportions. But the moment may be inauspicious, and what comes to mind can be rough and clumsy. The felicitous expressions remain beyond reach, and one's speech remains flat. Through writing, language turns into marble one can shape in outline, come back to, carve, refine, and polish. It becomes possible, as Nietzsche has it, "to work on a page of prose the way you work on a sculpture." [32]

Part Two

Cultural Information
Information for Reality

Producing Information
Writing and Structure

The Analysis of Structure

Natural information pivots on natural signs—clouds, smoke, tracks. Cultural information centers on conventional signs—letters and texts, lines and graphs, notes and scores. There is of course something like human culture before there is *cultural information* in the sense I use the words, and something like natural information remains pervasive and important after cultural information has arrived on the scene.[1] In the middle of Manhattan, darkening skies and sudden gusts alert you to a thunderstorm. The Empire State Building can help you to orient yourself. People gathering at an intersection indicate an unusual event. Still, the rise of cultural information marks the beginning of a new relation between humanity and reality. Human culture lay lightly or narrowly on nature until the vehicles of cultural information became available and aided in the moral and material transformation of the human condition, a development that has reached a crescendo since the industrial revolution.

While natural information is *about* reality, cultural information is distinctively *for* the shaping of reality. Cultural information, however, can be about, as well as for, reality. Architectural drawings or a musical score can be a record as easily as a recipe for making buildings or music. In the sketchbook of the medieval master mason Villard de Honnecourt, plans that inform us about existing churches are mingled with plans that Villard offered as instructions for future buildings (see fig. 3).[2] Gregorio Allegri wrote a score *for* the *Miserere* that was to be performed in the Sistine Chapel only and was therefore kept secret. When the fourteen-year-old Mozart, on having heard

FIGURE 3 Villard de Honnecourt's and Pierre de Corbie's proposal for a choir (top) and
(bottom) Villard's sketch of the choir of St. Estienne in Meaux (ca. 1235)

the psalm once, wrote down the score from memory, it was information *about* rather than for the performance of the music.

Regardless of whether cultural information is for or about reality, it differs importantly from natural information in the way it arises and comes to be present. Natural information emerges of itself, intimates rather than conveys its message, and disappears. Cultural information, to the contrary, is wrested and abstracted from reality, carries a definite content, and assumes an enduring shape. Before it can be encountered, it has to be produced by human hands. Yet even before it can be produced, the signs that are used to extract and convey information must have evolved.

The evolution of the alphabet has been meandering and playful, and its use becomes easy and natural to the competent practitioner. Thus the origin and employment of letters conceal the rigorous analytic achievement they represent. Plato, however, knew well that writing was as powerful a model as it was a vessel, and that it constituted a cultural force not only because of the contents it could convey, but also because of the structure it exhibited. Structure, of course, is crucial to information, and the search for structure is the quest for the secret of the nature of reference—the tie between signs and things. The driving force is the unspoken hope or belief that we can come to know the world clearly and comprehensively if we can penetrate the mysteries of structure, that is, uncover the ultimate constituents and the lawful arrangements of signs and things.

Though he was skeptical of writing, Plato was enthusiastic about letters. Writing he considered a threat to the primacy and vigor of community life. But the community Plato had in mind was the circle of his friends and students that met in the grove named after Akademos. In general, Plato was even more skeptical of community and orality than of writing. Athens, his native city, had during his youth engaged in disastrous foreign ventures and gotten entangled in a ruinous war with her ancient rival Sparta. Soon after, she condemned his teacher Socrates to death. While Plato singled out the Sophists as the major faction of irresponsibility, he also indicted the traditional oral culture for its failure to protect the good life against the corrosion of sophistry. Plato regarded Homer, the singer of songs and teacher of Hellas, as the emblematic narrator of a culture that was entangled in the shifting semblances and immediacy of life.[3] To the narrative attitude of Homer, Plato counterposed the analytic procedure of the philosopher.[4]

When Plato had Socrates, Philebus, and Protarchus analyze and define the good life, and when the infinite number of kinds of pleasure and wisdom threatened to confuse the discussion, the analysis of language served as the model for the clarification of the moral quandaries. In this particular piece, Theuth is the hero. He noticed, Plato says, that the infinity of speech sounds was structured, and so he distinguished its elements "until he knew the number of them and gave to each and all the name of letters."[5]

Moreover, Theuth recognized that letters have a special distinctiveness, one that today we call digital. For every mark on paper that is supposed to be a letter it must be clear *which* letter it represents. All the various *A*'s that have been and will be written are instances of the first letter of the alphabet. If we were lax about this and would routinely allow *A*'s to shade over onto *R*'s and *R*'s onto *P*'s and so on, we could never be sure of what letter a particular inscription represents, and writing would be overtaken by confusion and dissolution. Thus competent writers and readers are familiar with the model or type of each letter. As writers they fashion each mark or token according to these models or types. As readers, they match each token with its type.[6]

To be able to identify marks on paper this way, one has to know how each letter, in an alphabet of twenty-six capital letters, differs from all the other twenty-five letters. Every letter has a twenty-fivefold distinctiveness. An *R* differs from an *A* in having a rounded upper part. It differs from a *B* in having a straight line where the *B* has a lower roundedness. And so for all the remaining letters. Socrates must have had something like this in mind when he concluded his account of how Theuth discovered the digital nature of letters: "Perceiving, however, that none of us could learn any one of them alone by itself without learning them all and considering that this was a common bond which made them all one, he assigned to them all a single science and called it grammar."[7]

Alphabetic writing shows that language can be analyzed without remainder into a finite number of definite elements—spoken language into sounds, written language into letters. Plato furthermore thought that the analysis of language into sounds or letters reflected the reducibility of reality into elements.[8] "Element," in fact, is a word he uses for the ultimate constituents of both language and reality. What happens to the meaning of language and reality when they are brought down to their elementary levels? Plato's boldest answer was to the effect that meaning is the same as composition and that therefore objects devoid of composition, that is, the irreducible elements, have no meaning.[9]

Alphabetic writing certainly supports that view. A text considered only as regards its spelling, that is, merely as a sequence of letters, has no meaning, and disagreements of interpretation cannot arise. But that the world can similarly be reduced to such a fundamental level and at bottom has no meaning is more difficult to grasp, and some people find it impossible. Reality, to a poetical and scientifically untutored mind, contains distinct and meaningful objects that stand out against the indeterminate background of the world. Trees, birds, and boulders are prominent and clearly bounded things, and they are embedded in a dense and continuous context that inexhaustibly shades off into the distance. Trees are rooted in the soil and nourished by the rain. Birds are enveloped by the air and eventually vanish into the earth. Boulders detach themselves from the mountains and come to rest on the valley floor. Pictographic writing similarly lifts particular kinds of things out of the web of reality and language and sets them down distinctly in pictographs. But no matter how many hundreds or thousands of things, events, and signs were singled out and fixed in such writing, there remained a dense and inexhaustible context of further objects, processes, and relations.[10] Alphabetic writing, to the contrary, suggests that reality is structured all the way down, and at bottom is composed of a small number of meaningless, but well-defined, elements.

If there is a reassuring reply to the apprehension that the meaning of reality is in jeopardy unless it too goes all the way down, the answer must lie in the demonstration of how synthesizing meaningless elements into compounds reconstitutes meaning. Plato, inspired by the school of Pythagoras, tried to provide such a proof, proposing that from two kinds of right-angled triangles we can construct the

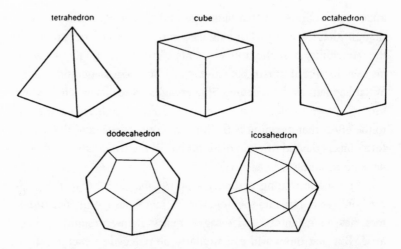

FIGURE 4   The regular polyhedra, also called the Platonic solids, from *The Penguin Dictionary of Mathematics*, ed. D. Nelson, 2d ed. (London: Penguin, 1998), 334. ©Penguin Books Ltd., 1989, 1998. By permission of Penguin Books Ltd.

regular solids—the tetrahedron, the cube, etc.; that four of these solids constitute the four traditional elements—fire being made up of tetrahedrons, earth of cubes, air of octahedrons, water of icosahedrons; and that the geometrical properties of the solids, such as pointedness in tetrahedrons, account for the meaningful properties of the elements, such as the penetrating force of fire (see fig. 4).[11] Working his way up to ordinary reality, Plato tried to show how all things, living and not, are composed of combinations of these elements. To the modern reader, Plato's attempt at deriving the florescence and abundance of reality from such spare components seems fanciful and arbitrary and casts doubt on the very claim that a meaningful world is constructible from meaningless elements.

We cannot be sure whether the analysis of language into sounds and letters in fact inspired Plato and his predecessors to propose an analysis of reality into elements or atoms. It seems likely, at any rate, that the alphabet promoted the particular analysis and synthesis of language we call grammar. Living language immediately encountered is so fluid, embodied, contextual, and evanescent as to discourage if not defy analysis. But once written down, especially in the parsimonious way of the alphabet, language appears stable, structured, and already analyzed into its ultimate constituents. Thus the task of synthesis seems clear and feasible. It is merely a matter, so it appears, of finding the rules of combination that will synthesize letters into words and words into sentences.[12]

Yet the plausible view, inspired by writing, that language constitutes a synthetic system of conventional signs is most likely in error. To be sure, language has its conventional and synthetic aspects. Logic and grammar have been the disciplines that have uncovered and explicated numerous formal structures exhibited by language. But a complete and consistent account of the structure of language has been an unreachable and even receding goal. The best efforts of recent linguistics unavailing, the rules of grammar have been vague when comprehensive and limited when precise.[13]

The moral we may have to accept is to the effect that the conventional and formal features of language are mere aspects of a natural and contingent phenomenon that, to be sure, is structured at the level of the molecules and neurons language is embodied in, but fails to exhibit exhaustive lawful structures at higher levels. In its living and deeper regards, language is an organism of natural signs and akin to the natural eloquence of things that projects and sustains natural signs.[14] Human eloquence, of course, is the most articulate and intricate. Still, it is as natural and intimate to the unself-conscious and competent speaker as commanding presence is to a landmark.

But the dream of discovering a thoroughly lawful structure in spoken and, a fortiori, written language was powerful in Plato and has remained so to this day. In the dialogue *Cratylus,* Plato has Socrates put the guiding conviction thus: "For the ancients gave language its existing composite character; and we, if we are to examine all these matters with scientific ability, must take it to pieces as they put it together and see whether the words, both the earliest and the later, are given systematically or not; for if they are strung together at haphazard, it is a poor, unmethodical performance, my dear Hermogenes."[15]

The dream of lawful structure, moreover, extends from language to reality and culminates in the hope that at bottom the structure of language and the structure of reality will be seen to coincide, the coincidence providing the tightest possible bond between signs and things. Information theory has never been able to uncover or forge this link. Modern information technology, however, has succeeded in devising a kind of equivalence between information and reality.

Plato toyed with this project, considering words where the quality of sounds seems to match the quality of the things they refer to.[16] But the rule of onomatopoeia does not go very far, and counterexamples are easily found. In the end, Plato recognized that contingency and convention, the apparent antagonists of structure and regularity, most often provide the answer, if in fact they constitute an answer, as to why a certain word refers to a certain thing.

The Synthesis of Structure

Though writing failed to reveal comprehensive structural rules of language and the Pythagorean program of constructing reality from the regular solids remained playful speculation, building pointed the way to real and general rules of construction. To be human is to mark out one's place in the world, and circles and squares were the first shapes people drew. During the long period of hunting and gathering, temporary or movable shelters, lean-tos, huts, and lodges were laid out in round or rectangular patterns.

When some 10,000 years ago the Neolithic revolution led to the construction of permanent dwellings, clay became the raw material par excellence, particularly in the Near East.[17] Clay was formed into bricks, as well as into pots, figurines, and tokens. The brick began as a lump of clay and was given one plane surface after another until in time it assumed its paradigmatic shape as a solid bounded by six rectangular surfaces.[18] Thus the large geometry of site and ground plans was joined by the small geometry of bricks and, not to be forgotten, by the yet smaller geometry of the clay tokens, tetrahedrons, cubes, cylinders, spheres, and cones.[19]

As alphabetic writing exemplifies the analytic side of structure, so building with bricks exhibits its synthetic side. The inventors of the alphabet had shown how all of language can be decomposed into twenty-six elements. Builders using bricks demonstrated how one kind of element can yield all kinds of structures—platforms, stairs, columns, walls, and vaults—and how these can be combined into sewers, farmsteads, residences, palaces, and temples.[20]

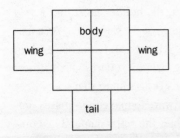

FIGURE 5   An agni (Vedic fire altar) in the shape of a falcon

The enterprise of locating, outlining, and erecting a holy place was the most eminent feat of construction in a premodern culture. The sacredness of a place came to be bound up with the accuracy and orientation of its shape. Placing stakes and stretching ropes carefully and punctually led to geometry. A ritual that in India at least until 1975 had stretched back in continuous tradition for some 3000 years is the Vedic construction of the agni, the fire altar.[21] The agnis were platforms, composed of five, ten, or fifteen layers of bricks and had the shape of circles, squares, rectangles, or of a falcon in flight. The ancient manuals for constructing the altars, the Śulvasūtras, tell us that for a certain predicament an altar of a certain shape was required. For example, one "who has enemies should pile the *agni* in the form of a chariot wheel."[22] The area for such a circular altar is prescribed and must be equivalent to an area given as a square. Thus one needs to circle a square, and the Śulvasūtras tell the supplicant how.[23]

In another instance, the sacrificer is instructed to build a falcon-shaped altar of seven unit squares: four squares for the body, one for the tail, and one each for the wings (see fig. 5).[24] In an ascending order of sacredness the next altar is to be built in the same shape, but one unit square larger and so on until the altar is one-hundred-and-one times as large as the first. Making an altar of seven unit squares into an eight-unit-square altar of the same shape requires the addition of one seventh of a unit square to each of the original seven squares. This can be accomplished by slicing the additional square into seven parts and adding one slice to each of the seven original squares.

Now the pious builder has an altar of the right area, but of the wrong shape because it is composed of seven rectangles instead of seven squares. To remove this blemish, the builder must find a way of converting each rectangle into a square of the same area. As a first step toward that goal, he slices the slice in two and attaches each of the thinner slices to one side of the original unit squares. That almost yields the desired square. The builder gets a square that has a little square missing in one corner. But if he were to make the square complete, he would get the wrong area albeit the right shape. The completed square is too large by the area of the little square (see fig. 6). What the builder needs is a square that is equal to the big (completed) square minus the little square:

desired square $=$ big square $-$ little square

Putting the same relation differently:

$$\text{desired}^2 = \text{big}^2 - \text{little}^2$$

And abbreviated:

$$d^2 = b^2 - l^2$$

which is equivalent to:

$$b^2 = d^2 + l^2$$

This is of course the Pythagorean theorem, and it holds the key to the final step for completing the agni builder's task. The theorem tells him that there is a right-angled triangle of sides b, d, and l. He has b and l, the sides of the big square and of the little square. Having these two sides and one angle (the right angle), the builder can construct the full triangle and so get the third side d, the side of the desired square (see fig. 7). This is roughly how the Śulvasūtras instruct the builder to proceed. They reflect a tradition that goes back to somewhere between 1000 and 800 B.C.E. Evidently the Vedic priests knew of the Pythagorean theorem several centuries before Pythagoras, who must have been a conduit of ancient geometry and arithmetic.

Just as the origin of counting and letters is likely to be traced back to sacred rituals, so is the beginning of geometry. The history of Greek geometry too shows traces of religion, references to gods, temples, and sacrifices.[25] But unlike the Vedic priests, the Greeks

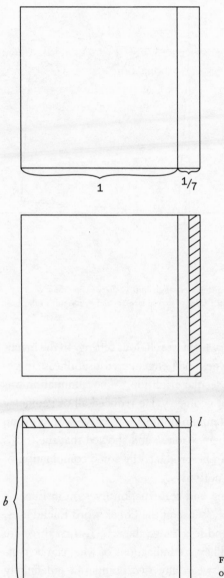

FIGURE 6 The first step
of converting a rectangle
into a square

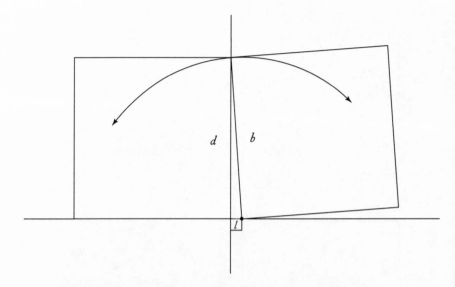

FIGURE 7   The second step of converting a rectangle into a square (using the Pythagorean theorem to obtain from the sides of two squares, b^2 and l^2, the side of the desired square, d^2)

eventually detached geometry from its religious setting. In the invention of the alphabet, playfulness had engendered usefulness. In geometry, sacred leisure led to secular erudition.[26] The culmination was Euclid's axiomatic system of geometry. He reduced all of geometry to five postulates and five common notions (today we would group them all together and call them axioms) and showed that the truths or theorems of geometry can be produced by some combination of the postulates and common notions.

The language of building and construction lives on in Euclid's geometry. To construct, in the sense of the Greek word Euclid uses, is to place things together and to arrange them.[27] His first three are postulates of construction. They are indications of what can be constructed—a straight line between any two points, an indefinitely long extension of the line beyond one of the points, a circle with any point as center and any length as radius.

The word construction that is used in Greek for buildings and furniture also designates a formal step in Euclid's demonstrations of the geometrical propositions.[28] Thus to demonstrate the Pythagorean theorem, Euclid has you consider a right triangle and then construct squares on the three sides.[29] In addition he has you draw five auxiliary lines within the figure of triangle and squares. The remainder of the demonstration consists of pointing out the equality of angles, triangles, rectangles, and finally of the areas of "the squares on the sides containing the right angle" and "the square on the side subtending the right angle."[30]

The epochal achievement of Euclid's geometry, however, lies not in its ability to illuminate a particular mathematical or physical construction but in apparently capturing the very structure of space. As Gordon Brittan has put it, "geometry is the form of the world," and to have all the truths of geometry condensed into five postulates and five common notions is to have captured, so it seems, the essence of the world's *form* and one half of the structure (the other half being the *content*) that needed to be laid bare if the workings of reference are to be understood.[31] As Brittan has shown, however, it takes persistence and boldness to grasp mathematical structures in their true generality.[32]

To the ancient Greeks, geometry and arithmetic were separate fields, and construction and computation were entirely different operations.[33] They would have regarded the fusion of the geometric and algebraic realms and of the constructive and computational methods that the philosopher René Descartes (1596–1650) achieved early in the seventeenth century as entirely wrongheaded.[34] Rigorous knowledge, so they thought, requires careful distinctions. The pebbles in your hand have a number, a so-many, whereas the wine in your cup is a magnitude, a so-much.[35] Numbers can be added while magnitudes can only be compared to each other. A circle, to the Greeks, was like a cup of wine, a shape and magnitude. You can ask how a cup of wine compares to a goblet of wine or a circle to an ellipse.[36] But a circle is not specifiable as a set of numbers just as a cup of wine is not a collection of wine units. To combine numbers and magnitudes is to be badly confused and must be prohibited.

On the Euclidean plane, Descartes imposed a grid, defined by a vertical reference line, the y-axis, and a horizontal one, the x-axis. Within this grid, a circle is defined by the equation:

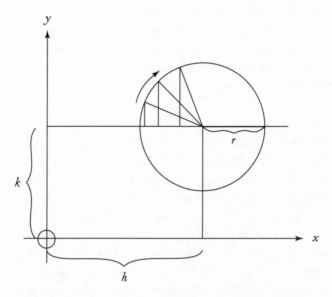

FIGURE 8　A circle in a Cartesian coordinate system being swept by the hypotenuse (the longest side) of right triangles

$$(x - h)^2 + (y - k)^2 = r^2$$

To the untutored in analytic geometry this looks, if anything, like the Pythagorean theorem. In fact one can think of the equation as describing infinitely many different right triangles. Now imagine them succeeding one another in the shapes prescribed by the equation, and the long side of these triangles will sweep the area of a circle on the coordinate grid as the second hand does on the face of a clock (see fig. 8).

To get an actual circle, however, units of whatever length need to be marked off on the baseline, the x-axis, and on the principal meridian, the y-axis, of the coordinate system. Numbers have to be picked for h and k—for example, 10 and 7— that will determine the location of the center of the circle on the grid, and a number for r— say 5—that will fix the radius of the circle. Then any two numbers for x and y that fit the equation

$$(x - 10)^2 + (y - 7)^2 = 25$$

　　　　　　　　　　　　　　　　　　　　　　CHAPTER SIX

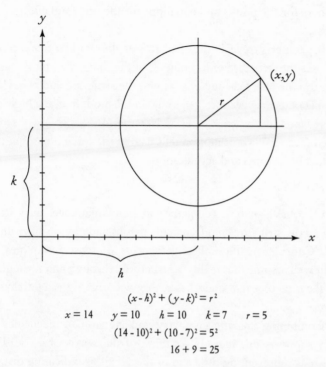

$$(x-h)^2 + (y-k)^2 = r^2$$
$$x = 14 \quad y = 10 \quad h = 10 \quad k = 7 \quad r = 5$$
$$(14-10)^2 + (10-7)^2 = 5^2$$
$$16 + 9 = 25$$

FIGURE 9   A particular circle in a Cartesian coordinate system with its particular values

will describe a point on the circle of the size and location indicated above (see fig. 9).

In this way magnitude and number have been joined. The shape, size, and position of a circle in general are captured by the algebraic equation that contains only generalized numbers, that is, letters. If you replace the letters with numbers, you obtain a particular circle whose properties and relations are expressed in numbers. These numbers in turn can be used to determine and compute the relation of the circle to anything else that can be specified and located in the coordinate system. Thus the coordinate system reveals a deep connection between magnitude and number, between the "so-much" and the "so-many."[37] The plane of the coordinate system supplies the space for figures, magnitudes, relations, and proportions. The coordinate axes impose a grid on that space and provide pairs of numbers for the points and equations for the lines and curves of the figures.

If geometry reveals the structure of the world's form, physics discloses the structure of its content, of the things that fill the world. Here too the task was one of grasping the structure and behavior of tangible objects ever more generally. In the modern era, a high point was reached in the work of Isaac Newton (1642–1727) whose celebrated (second) law of motion tells us that the force of a thing is the product of its mass and acceleration:

$$F = m \cdot a$$

Of course, analytic geometry and Newtonian mechanics are just some of the structures modern mathematics and physics have uncovered. Considering the collective success of these disciplines, one might well assume that reality is structured through and through and that the structure that underlies reality will give signs a purchase on things.

Combining analytic geometry and Newtonian mechanics certainly supports this idea. If we think of the x-axis of a coordinate system as representing time and of the y-axis as indicating distance, we can capture and picture the lawfulness of a body moving at a constant velocity by a line, diagonal to the axes and steeper if the velocity is greater. If the line is flat, the body is at rest. If the line begins to curve up ever more steeply, the body is accelerating. The intersection of a curve and a line would define the point at which a constantly moving body and an accelerating body have covered the same distance in the same time.

Producing Information
Measures and Grids

Grids of Space

When it appeared that language and reality were more like clocks than clouds, that they had a lawful structure composed of definite parts, it seemed reasonable to look for a tight structural fit between signs and things. Plato had hoped to see the structure of words mesh with the structure of things, a hope that was revived often, most recently in Wittgenstein's *Tractatus Logico-Philosophicus* of 1921.[1] The idea was that sameness of structure would allow language and reality to engage one another like cogs so that the movement of one wheel would faithfully reflect the movement of the other. If this conjecture had been substantiated, we would not only have perfect insight into the linkage between signs and things, we could also hope to repair and improve the linkage where neglect or abuse had allowed the gears of language and reality to be damaged. We could then obtain perfectly articulate and reliable information about reality as a matter of course.[2]

Though the structure and operation of language are far from fully understood, the advances of recent linguistics have made one thing clear. The more we have learned about language, the less its structures have come to look like portraits of the speaker or pictures of reality.[3] And accordingly, when language informs us about reality, it does not picture what is remote in time, space, or conception, but reminds us of the eloquence of things or prompts us to imagine their voice. Though such eloquence emerges from the physical structure of persons and things, it does so in contingent and unforethinkable ways.

To be sure, there are instances where a realm of reality exhibits broad structural features that can be pictured in a model or a graph as is true, for example, of celestial mechanics. And a portion of reality may be structured all the way up so that diagrams and formulas can pretty well disclose all there is to know as is true of a gallon of distilled water or a pure silicon crystal wafer. In such cases, something like structural sameness of signs and reality does in fact produce information about an aspect of the normally concealed essence of some thing.

Though all of reality is structured all the way down, most of it is not structured all the way up. The microscopic structures of physics and chemistry are rarely compounded into ordinary things according to rules as elegant as those of the natural sciences. And the value of summary scientific information such as the weight of a person or the height of a building is limited. There is, then, an information gap between the structural information that is uncovered by scientific analysis and measurement and the contingent information about the expressive faces and eloquent voices of people and things. The gap is large and contains everything in space and time that is unremarkable and irregular. How can we obtain information about that?

As it turns out, information can be produced by structure imposed as well as by structure revealed or eloquence conveyed. Eventually the extraction of information from reality by means of structural devices not only covered the information gap but became a universal instrument that enhanced science, overtook art, and has come to capture everything. The imposition of structure began gradually by fixing measures for distances. Such measures were taken from the ways humans touch and experience reality—hence the span, the hand, and the foot, the moon, the day, and the hour. From there further measures developed, such as the distance one can walk in a certain time. And eventually measures were reconciled with one another, twelve inches to the foot, three feet to the yard, five and a half yards to the rod, three hundred and twenty rods to the mile.

Linear measures, however, are a weak imposition on reality and produce limited information. The archetypal instrument for the extraction of information from reality is the grid. For theorists it is a powerful metaphor to illuminate the production of information. Talking about a person's ability to carve the world into the individual things (to "individuate" them) that constitute the objects of information, Keith Devlin explains: "Picture the agent's individuation mechanism as consisting of a family of *grids* through which the agent can 'view' an otherwise indiscernible world. These grids pick out (or determine) the individuals, relations, locations, etc. that constitute the ontology of our theory (for that agent)."[4] Diana Raffman similarly suggests that we think of the relation between the ineffable plenitude of sounds (the "literal pitch stream," or LPS) and the division of the continuum of pitch into the familiar twelve intervals of the chromatic-diatonic scale (or CD) we can clearly refer to as follows: "You can think of the interval schema as a kind of template or grid through which the LPS is passed. The grid is only as fine-grained as the CD-pitches—A-natural, B-flat, B-natural, and so on—so that the incoming stimuli are categorized only that finely."[5]

The first trace of an actual and mundane grid was the reference line that the philosopher Dicaearcus, a follower of Aristotle and contemporary of Euclid, drew across a map of the ancient world, beginning in the east at the straits of Gibraltar, the "Pillars of Hercules," through the Mediterranean basin and across Asia Minor to the Indian or "Eastern" Ocean.[6] Eratosthenes added a meridian, anchored in Alexandria, and additional lines to produce a rough grid. In the second century C.E., Ptolemy produced a regular grid that reflected the earth's spherical surface. It amounted to the imposition of Euclidean lines and arcs on what Ptolemy took to be the habitable world.[7]

Grids came truly into their own, however, when the eloquence of reality began to decline and a burst of collective curiosity broke through the boundaries of the ancient and medieval world. Grids wrested reliability from contingency and produced information that made reality not just perspicuous but surveyable. Trees, creeks, rocks, and caves can eloquently define a piece of ground, but they do so precariously, given the winds of change. Consider the description of a ten-acre tract, "The thirty-first lot in the Western Division of Common Rights in the Town of Manchester, made in the year of our Lord one thousand six hundred ninety-nine":

At the northeast corner with a maple tree between him (Captain West) and Abraham Masters, from that westerly 30 poles to a hemlock tree between him (Captain West) and Abraham Masters, from that southerly westerly 39 poles to Morgan's Stump, from that southeasterly 44 poles on said West's farm line to a black oak, and from that 66 poles northwestward to the first point.[8]

The account is not much more elaborate than the one in Genesis that describes Sarah's and Abraham's burial place in Hebron, and if the information above were all we had to go on, the thirty-first lot would be as difficult to locate as Ephron's field and cave. Trees die, creeks change their course, rocks get moved, and so accordingly does the information they convey. Compare the description above with the "legal description" of the ten acres where the Salish used to have their camp:

Southwest One-Quarter Southwest One-Quarter Southwest One-Quarter, Section 25, Township 14 North, Range 19 West, Montana.

The description is anchored in a coordinate system that divides Montana into squares, six miles to the side, townships so called. To locate the vicinity of the Salish campsite you move up fourteen townships north from the x-axis or baseline and nineteen townships west from the y-axis or principal meridian. The square miles of each township are numbered one through thirty-six, beginning in the northeast corner and moving left to the northwest corner, dropping one line down, numbering from west to east, meandering back and forth in a kind of sequence that is at odds with the way we write now though it follows an archaic direction of writing that moved as an ox does when it pulls a plough.[9] Section 25, then, is the rightmost square mile of the second to the last line of its township, and moving to the southwest quarter of section 25, then to the southwest quarter of that quarter and once more the next smaller southwest quarter, you have at last located a square of ten acres.

This coordinate system goes back to the federal Ordinance of 1785. Jefferson, ever a man of the Enlightenment, is credited with this rational and geometrical system of surveying and delineating the territory north and west of the Ohio River that in the 1780s had passed into the hands of the federal government. In Montana, the

CHAPTER SEVEN

origin of the coordinate system, the initial point where the Montana principal meridian and base line intersect, lies ten miles southwest of the origin of the Missouri—the confluence of the Jefferson, Madison, and Gallatin—and fifteen miles west southwest of Manhattan, Montana.[10] Manhattan in New York City has, like most cities, a coordinate system of its own. Its initial point lies at the intersection of 1st Street and 1st Avenue, just northwest of downtown. Manhattan is a microcosm of this country, the irregular network of streets in downtown Manhattan reflecting the intimate fabric of metes and bounds of the colonial states and the regular grid of midtown and uptown corresponding to Jefferson's rational system in the Public Domain of the midwestern and western states.

The Montana principal meridian and baseline are themselves fixed in relation to the global coordinate system of longitudes and latitudes. Hence the vicinity of the Salish camp is precisely and permanently fixed by the description above as long as the structure of the interlocking grids remains known. Since the equator had traditionally been accepted as the global baseline, only the determination of the prime meridian needed to be agreed upon. After much nationalist wrangling, "the center of the transit instrument at the Observatory of Greenwich" in England was in 1884 accepted as the fixed point of the international prime meridian.[11]

Precision and reliability are not the only benefits of uniformity and commensurability. The latter seem to compare poorly with the envelopment of a place where distances are measured in happy skips, busy steps, or solemn paces. To be enveloped, however, is not always to be embraced. The envelope of a familiar place can stifle curiosity and inspire fear of the wider world as well. The medievals, whose world was centered around Jerusalem and well-ordered in all directions as Dante teaches us, felt forever surrounded and threatened by the distant and dimly known Gog and Magog, those cruel and ferocious tribes that Alexander the Great had driven off for who knows how long.[12]

Similarly, the familiar and meaningful world of the Salish, when they camped in the Rattlesnake Valley, was surrounded by large areas that were insignificant, all but inaccessible, and barely known—steep and trackless side canyons, treacherous slides and rock walls, dry and barren ridges. Such, at any rate, are the vague and forbidding margins for someone who today, unaided by a map, follows the signs and hiking trails of the Rattlesnake. If you carry a topographic map of the valley and are a proficient reader of maps and territory, the cartographic information will lift the veil of ignorance and open up the lay of the land. Envelopment yields to perspicuity and more—to surveyability. So it did on a global scale when in the sixteenth century advances in cartography and navigation led to the voyages of discovery and revealed the world entire. And so it did once again in the second half of the nineteenth century on a more limited and intensive scale when the U.S. government sent out contingents of the army to explore and survey possible routes for railroads to open up the western territories for white settlements and to link the Atlantic to the Pacific coast.[13]

Measures of Time

What grids did for space, the clock did for time. In fact, the coordinate system of latitudes and longitudes was only half helpful until a reliable clock had been invented. Sailors at sea can determine their distance from one reference line, the equator, by the angle of the sun at noon. But that angle provides no information about their distance from the other axis, the longitude that runs through the observatory at Greenwich. To know how far west of Greenwich they were, navigators had to know how much later their moment of noon was than the moment of noon at Greenwich. If on departing from England they took a chronometer set at Greenwich time, it was easy to calculate the delay of noon time and the spatial distance from Greenwich.[14]

The first mechanical clocks were designed in the thirteenth century, and so was the abstract grid of sixty-minute hours and sixty-second minutes.[15] But a clock, indifferent enough to rocking and sufficiently impervious to changes of temperature to keep accurate time at sea, was not available until 1759 when John Harrison finished his magnificent chronometer no. 4.[16] Determining longitude at sea had been a vexing problem. The inability of sailors to plot their position precisely had led to untold miseries of confusion, disease, starvation, and death.[17] In the eighteenth century, a reliable clock was by far the simplest solution to the problem. Yet Harrison's breakthrough encountered massive resistance in the face of an alternative solution—the lunar distance method. What endeared the latter to scientists and sailors was its traditional cast and the intimacy with astronomy and mathematics it required. An accurate determination of the moon's position in relation to the stars plus extensive calculations could yield an accurate location of longitude.[18] In comparison, as Dava Sobel has it, "this device of Harrison's had all the complexity of the longitude problem already hardwired into its works. The user didn't have to master math or astronomy or gain experience to make it go. Something unseemly attended the sea clock, in the eyes of scientists and celestial navigators. Something facile. Something flukish."[19]

Here is an early example of how the progress of information technology yields information more instantaneously and easily while at the same time it disengages us from reality and diminishes our expertise, the latter being assumed by the machinery of a device. It is an example too of how sound and helpful such a device can be. Its underlying pattern, moreover, has proven a powerful tendency, and even the chief advocate of the lunar distance method could not escape it. Nevil Maskelyne sought to lighten the burden of the method. As Sobel tells us, "[b]y incorporating a wealth of prefigured data, he would reduce the number of arithmetical calculations the individual had to make, and thereby dramatically shorten the time required to arrive at a position—from four hours to about thirty minutes."[20]

But long before Harrison had constructed his punctilious device, people in the European cities had learned to fit the course of their lives into the grid of hours and to coordinate their activities accordingly. The charting of time too led from variety to uniformity. Prior to clock time, hours and days were longer in the summer than in the winter.[21] A day's work was measured not by hours but by what needed to be done. Clock time equalized the days and fixed the working hours. It introduced precision, replacing midmorning with 10:30 A.M. And it made the implicit rhythm of life's daily round evident and surveyable. But the employment of clock time also shows that the ability to see things in a new way can lead to ordering them in a new way as well. Abstract information becomes concrete transformation.

Lines of Print

One of the finest illustrations of the actual use of a literal grid is Albrecht Dürer's (1471–1528) woodcut of the draftsman who looks at a reclining nude through a grid, while steadying his gaze with an eye-piece, and transfers what is captured by the grid onto paper that is properly squared to receive and convey the information obtained by the draftsman (fig. 10). The woodcut, moreover, is rendered as though Dürer himself had regarded and fixed the scene through a grid. It is information on how to obtain information.

Dürer drew the design for his woodcut early in the sixteenth century when the eloquence of the fine arts was beginning to fall silent in part through the iconoclasm of the Reformation. The woodcut, part of Dürer's *Instruction for Measurement with Compass and Straight Edge*, was not only an instruction about the use of grids but also an instance of the most extensive and influential grid to be imposed on reality—the printed page.[22]

For practical reasons, writing space itself has always required some coordinate system. In speech, the sounds that are reflected in letters or logographs are ordered by the flow of time. On a two-dimensional surface, an initial point and a coordinate convention need to be fixed to determine the direction and arrangement of lines. The history of writing has seen all kinds of rectangular coordinate systems with all sorts of initial points and directions of writing, and there has been at least one instance of a polar coordinate system on a disc where writing starts at the center and spirals outward clockwise.[23] Our present system of writing in horizontal lines from left to right, beginning in the upper left corner, was established around 500 B.C.E. in Greece.[24] The orderliness and accessibility of writing were heightened when around 200 C.E. the scroll was replaced by the codex. Though writing on a scroll was arranged in columns or pages, there was no random access to a particular page. Just as getting to the middle part of an audio or video tape requires you to wind your

FIGURE 10   Albrecht Dürer, *Draftsman Drawing a Reclining Nude*

way to it, you had to unroll one side of a scroll and roll up the other to locate a particular page. A codex is a stack of pages sewn together on one side to form what we now call a book.[25] Finding particular information on a particular page became easier in a codex and became easier still when in the early Middle Ages indexes and tables of contents began to serve as guides to numbered pages and layout in paragraphs as a guide to a particular passage on a page.[26]

Writing as an orderly grid of information grew more powerful by orders of magnitude through printing. The printed page has a rigor and universality that make a manuscript appear to be frail and hidebound. Compared with oral transmission, of course, writing is the very model of precision and reliability.[27] It detaches information from the fallibility of memory and gives it the distinctiveness and robustness of digital signs. Digitality would allow for faultless copying were it not for the foibles of scribes who commit errors of oversight, misunderstanding, and unwarranted emendation.[28] Manuscript copies of expansive writings, moreover, were laborious to make and relatively few in number. Thus writing, though so much more mobile than a monumental sign, remained rather place-bound, tied to the library of a prince, a monastery, or a cathedral school. Scholars had to travel to a particular site if they wanted to consult a particular work.

Printing provided a rigid and general framework for information. Unlike copying, typesetting was a labor that bore fruit in hundreds or thousands of copies. Hence it warranted much greater care and precision. And once a printed book was disseminated, one and the same text was before thousands of eyes who could check and compare it and send corrections and additions to the printer for incorporation in subsequent editions. Printing made it possible to extract an authoritative and generally available text from scattered and corrupt manuscripts.

What is true of texts holds equally for maps. A particular place such as the monastery Klosterneuburg near Vienna could serve as a collection point for charts, maps, and geographical reports. But to copy a map by hand is to make it less accurate. A copy of a copy would be less precise yet until at length the multiply descendant copy of a map would be as misleading as helpful. What typography did for texts, xylography and engraving did for maps and illustrations. They justified and rewarded careful work and invited and channeled improvements from the users of the maps and charts to the printer. Printing made it feasible and sensible to resurrect Ptolemy's world map from his writings.[29] A comparison of his map with the very rudimentary one that had come directly out of medieval geography and was the first to be printed demonstrates that without printing there is no universal coordinate system to serve as a framework of improvements.[30]

More even than texts and maps, mathematical and astronomical tables prospered in print. To compute logarithms or the positions of planets is a laborious task. Checking tables for accuracy is tedious. Giving them to a scribe for copying is to court despair because some of the labor of computation and compilation is bound to be tainted by errors and carelessness.[31] Here again care of typesetting and proofreading and emendation of later editions provided more precise, powerful, and widely available systems of information. They were far from perfect, to be sure. In the early nineteenth century, Charles Babbage became both intimate and frustrated with the thousands of errors contained in printed mathematical tables and was moved to exclaim, "I wish these calculations had been executed by steam!" A year later, in 1822, he built his first mechanical calculator.[32]

Even so, compared with manuscripts and drawings, print created an entirely new medium of information. It liberated texts, maps, and tables from their unique incarnations and projected them on a universal screen that was accessible to many people at once in many places and at any time. It was a surface that in our time has unfolded a third dimension and has come to be known as cyberspace.

Just when printing in the early sixteenth century established itself as the dominant coordinate system of information, another grid for fixing a realm of reality achieved its final form. Musical notation as we know it today is a coordinate system that allows us to plot pitch and duration over time. As with alphabetic writing, the relatively clean and rational structure of contemporary staff notation is the result of a long history of experimentation and contingency that ended with the convergence of naming and symbolizing groups of notes in sequential writing with the practice of placing notes on a picture of the strings or the keyboard of an instrument.

If the printed page was the most extensive means of gathering information from reality, money was the most intensive. It is the common measure of things par excellence. As Aristotle saw, money tells us how to relate a house to pairs of shoes and the builder's to the shoemaker's time.[33] Its rise in the west parallels that of the clock and the printed page. Like the former, the money economy began its reign in the thirteenth century, and like the latter, it had covered Europe with a reasonably precise and perspicuous grid of measuring goods and services by the sixteenth century.

Money, being the most powerful instrument of commensuration, has also most clearly revealed the resistance of reality to the imposition of grids and measures. The progress of money will have reached its goal when one and the same currency measures everything everywhere. But national reservations have so far stood in the way of strict uniformity of currency, and moral concerns have limited its universality. To agree on a global currency, countries would have to give up their ability to direct the domestic vis-à-vis the international economy, restraining, for example, imports and spurring exports by devaluing the national currency or cooling an overheating economy by restricting the money supply. Moral apprehensions have moved us to protect certain things from being officially measured in money and available for money, things such as human lives, justice, love, health, education, wilderness, and legislation.

Clearly these limitations are under pressure from the intrinsic tendency of money toward uniformity and universality. Those who support this tendency often do so under the banner of fearless enlightenment that has flown above the earlier fighters for uniform, precise, and perspicuous information about reality. It is true that the lack of one global currency conceals from us valuable information about the real differences in productivity and affluence in the various parts of the globe. Our unwillingness to move anything and everything into the market prevents strict comparisons between things that after all are parts of one and the same reality and influence each other whether we acknowledge it or not. What is the value of sexual attractiveness? How much does an acquittal cost? At the same time, money shows, more clearly even than clock time, that information about reality leads to the transformation of reality.

Realizing Information
Reading

Fundamentals

Cultural information allows us to venture beyond the actual into the possible and to set down in letters, lines, or notes what otherwise would remain a remote or elusive possibility. Writing and drawing convert information into a material that writers, composers, or builders can use to articulate their conceptions at leisure and at length. Thus, cultural information raises to a much higher level a distinctively human ability—to think before one acts. On paper we can try out and spell out in endless variation and detail what in an oral society needs to be risked or left untried.

Once the merely possible has become an actual design, a constructive kind of possibility arises. Books, scores, and plans are primarily signs, referring to possible things. Signs are vehicles and vectors. The meaning they convey directs us beyond themselves to things. They are instructions for the construction of some reality— a new vision of the world, a structuring of time, or a transformation of space. At their best, these changes enrich the world morally, practically, and tangibly. They raise our prosperity.

The process of converting the instructions of a design into something real I call *realizing information*. The paradigmatic kinds of such realization are reading, performing, and building. Realizing information is to take an abstract design and to have it come to life in the concrete world. This can be an exhilarating as much as a perilous event. A grand conception becomes a reality; yet no conception can entirely anticipate and control the conditions of its realization. The contingency of the world, its vagaries and frailties as well as its vigor and resourcefulness, assert themselves in unsurpassable ways. Realization of information is always the encounter, and sometimes the struggle, of the structure of a design with the contingency of reality.

It is a challenge we meet most often and inconspicuously in reading. The word "reading," in fact, means something like realizing information, to make sense of signs, broadly "to puzzle out" as when we read people's faces for clues to their mood or the water of a river to find a path for a canoe. More narrowly and normally, however, it is letters that we read and try to make sense of. Like all cultural information, writing continues and potentiates the function of natural signs—to provide information about reality. This is what records and reports do, and when we read them, the path of realizing information is relatively straightforward—it is a matter of remembering or imagining something that was or is already real. Straightforward relative to what? In relation to reading poetry or fiction where the reader is instructed to construct a new kind of reality.

Reading of whatever sort is a many-storied skill, both in the sense that you must read many stories to acquire it and in the sense that it is composed of many layers. The physiological and literal stories we understand fairly well. But as reading rises to its full realization, much of its structure is veiled by ignorance.

To begin at the lowest physiological level of perception, what we practice and experience as the smooth gliding of our glance along the lines of letters is, microscopically considered, a series of overlapping jumps, so-called saccades, and fixations on words, each a matter of milliseconds.[1] Saccades and fixations are native to all seeing and need not be learned. They must be adapted to reading, and practice is the only teacher for that task.[2]

It is otherwise at the level of converting letters into sounds and integrating sounds into words. Teaching has a time-honored place in the acquisition of this skill. Acquisition is particularly difficult in English since its orthography carries so much etymology. In many cases, what we see is Middle English, but what we say is Modern English. Hence the relation of letters to sounds is complex. The first task is to group letters into graphemes of one or more letters so that each grapheme corresponds to one phoneme or sound. The how of such parsing is perfectly achieved by a fluent reader, but hardly understood by the theorist. In "shepherd" *p* and *h* belong to different graphemes, in "periphery" they belong to one and the same. And of course one and the same grapheme can correspond to very different sounds. Consider *gh* in "rough" and in "ghost." Setting aside rare or unpredictable cases, there are 100 to 200 grapheme to phoneme correspondence rules—what amounts to a reversion from alphabetic to syllabic writing.[3] Finally, the reader must integrate the sounds into words and words into sentences that exhibit the usual patterns of intonation. How do readers get from the signs to the message, from letters to meaning? They follow a path that is a braid of many trails. In large part readers transform graphemes into phonemes, that is, groups of letters into sounds, imagined sounds when reading is silent. Once a letter has become sound, the oral comprehension of language can take over. In part, however, written and printed words have a unified physiognomy and are read directly and as units, much like logographs.[4] For the beginner it is in fact easier to take words as meaningful units rather than phonetic compounds.[5] But proficient reading requires analysis and synthesis, decomposition of a string of writing into letters and recombination of letters into a coherent word.

But reading is more than recognizing strings of letters as words and knowing their acoustic shape. Really to read is to comprehend. The scholarly examination of reading up to the level of word recognition is voluminous and detailed. Where it proceeds from word recognition to text comprehension it chiefly reveals perplexity and ignorance.[6] Of course reading and listening require some of the same skills, for example, the skill of parsing—the ability to group words together meaningfully. That we do this effortlessly does not mean that we do it mechanically. No one has any difficulty in properly parsing "Time flies like an arrow." But a computer, proceeding mechanically, was undecided between the normal and four other possible parses. The funniest of these alternatives can be suggested by the addition of an introductory phrase: "Fruit flies like bananas, time flies like an arrow."[7]

Comprehension

Whatever the complexities of parsing, it is clear that the context helps to resolve the structural ambiguities of sentences. Supplying a rudimentary framework by way of an introduction prompts the reader to settle on a particular parsing. But such prompting, though solving the contextual problem in one sense, aggravates it in another. The introduction, after all, is one more piece of information, beset by its own ambiguities and requiring yet another context to acquire a definite meaning. This observation highlights the difference between listening and reading that so distressed Plato. When we listen to someone speak, there is not only the pace and the rhythm, the inflection and intonation, the posture and expression of the speaker to help us make sense of language. There is as well a shared moral and material context to supply direction and orientation.

This context has typically been stripped away through writing, and it is the reader's task to provide it. We may think of intelligent reading as restoring to written SIGNS about THINGS the natural structure of information where, INTELLIGENCE provided, a PERSON is informed by some SIGN about a THING within a certain CONTEXT.

The reader is of course that person, and her intelligence allows her to construct a context for the signs to make coherent sense. Intelligence in this case is both the general skill of decoding signs and the fund of the particular experiences that are needed for the words in question.[8] Intelligence can construct a context of meaning because it previously has drawn on the context of reality. The significance of things, as the medieval philosopher Hugh of St. Victor (1096–1141) tells us, has priority over the significance of words, and the significance of things in turn arises from divine eloquence: "The philosopher knows only the significance of words, but the significance of things is far more excellent than that of words, because the latter was established by usage, but Nature dictated the former. The latter is the voice of men, the former the voice of God speaking to men. The latter, once uttered, perishes; the former, once created, subsists."[9]

In the same spirit, to cure Phaedrus of his infatuation with letters, Socrates, the inveterate urbanite no less, takes him out of the city into the world of natural things to remind him of the eloquence of the oak and the rock.[10] He charges Phaedrus to convey to Lysias, writer of flimsy speeches, the words of the nymphs who inhabit the fountain and the sacred grove, and he concludes his therapy with a prayer to "Pan and all ye other gods of this place."[11]

A powerful reading of a text requires that a reader has had vigorous experiences or at least harbors profound aspirations. What readers bring to a text varies greatly from epoch to epoch and from person to person. Therefore strong readings have little generality. This is widely acknowledged today and frequently overrated. Texts and cultural information have levels of generality that, although weak in meaning, are indispensable for the prospering of a culture. There is an inverse relation between the generality and the strength of textual meaning. At a time when momentous readings of texts such as the Bible or the Constitution are anything but widely shared, it is important to appreciate the shared if ambiguous layers of writing.

A text is most ambiguous at the level of spelling, and here we can get the widest, if not incontestable, agreement on the identity of a novel or a poem.[12] Abiding by the common agreement on the digital distinction of letters and the conventions of orthography is crucial to the integrity of higher and stronger readings. So is the next semantic layer, the level of literal or lexical meaning. Every word functions or refers in a general and typical way, carefully recorded in dictionaries. Lexical definitions are bland of course. The entry for *river* does not capture what Norman Maclean has gathered in "A River Runs Through It."[13] But it does assemble for us what the Blackfoot, the Rhine, and the Jordan River have in common. Lofty readings are likely to collapse if the fundamental level of literal meaning is neglected.

Teachers of reading have stressed this order of meaning since antiquity. *History* is their term for the literal and fundamental meaning of a text. Its importance was extolled in the twelfth century by Hugh of St. Victor, who, quoting and adapting Gregory the Great, said, "As you are about to build, therefore, 'lay first the foundation of history; next, by pursuing the "typical" meaning, build up a structure in your mind to be a fortress of faith. Last of all, however, through the loveliness of morality, paint the structure over as with the most beautiful of colors.'"[14]

In the mundane reading of reports, records, and recipes, the path from the literal to the full realization of a text is fairly straight and short. Sturdy and public contexts are available for the placement and invigoration of signs about things. There may have been a time when such contexts too subtended signs about the most momentous matters. The trust in such a context is reflected in the legend that when in the third century B.C.E. a Greek translation of the Hebrew Scriptures was requested by scholars in Alexandria, a delegation of seventy-two translators was sent from Jerusalem, and each produced verbatim the same Greek text, henceforth called the Septuagint (*seventy* in Latin).[15]

In fact for most of its history reading used to take place in a common moral and tangible setting. Until the end of the eighteenth century, reading was normally done aloud and to a company of listeners and mumbled if done by oneself.[16] In ancient times, silent reading was a rare and astounding skill. Augustine in the fourth century C.E. was amazed to find it in his teacher Ambrose: "But when he was reading," Augustine reports, "his eyes were led across the pages and his heart uncovered insight; his voice and tongue, however, were silent."[17] Not until the Middle Ages did silent reading become a well-established skill.[18]

To those who cannot read silently, the mere sight of letters and words yields no understanding. They must hear the words to comprehend them, just as amateur musicians must play a score to find out what the music sounds like. Where today, when we have difficulty in following someone who is reading to us, we are likely to say "let me see that," medieval authors who were doubtful about a passage they had dictated would say to their scribes "let me hear that," a practice that has come down to us in *audit*, our term for careful examination.[19]

Loud or mumbled reading seems slow and awkward to us, and so it was. But it was more than that. It engaged the body more intensively and impressed itself on memory more profoundly than silent reading. In Jewish and Muslim schools, bodily engagement and the capacity of memory are employed more fully through chanting and gentle rocking and swaying.[20]

In some places, common reading survived into this century. Consider the testimony of Norman Maclean about his youth in Missoula, Montana: "After breakfast and again after what was called supper, my father read to us from the Bible or from some religious poet such as Wordsworth; then we knelt by our chairs while my father prayed. My father read beautifully. He avoided the homiletic sing-song most ministers fall into when they look inside the Bible or edge up to poetry, but my father overread poetry a little so that none of us, including him, could miss the music."[21] By now the practice has died. "I need hardly tell you," Maclean notes sorrowfully, "that families no longer read to each other. I am sure it leaves a sound-gap in family life."[22]

Today, reading is overwhelmingly silent and private. But the cultural losses we have suffered when communal reading disappeared should not blind us to the charms and graces of today's typical reading. These virtues are most prominent when reading is poetical rather than utilitarian, when what is read are not records or recipes, but scriptures, novels, and poems.

Intimacy and privacy are the hallmarks of today's reading.[23] No sound comes between the reader and the text, no listener can intrude into the reader's comprehension. Though various contraptions and devices have been invented to disburden readers, the typical reader has been found in the same position for centuries—seated, holding the book, and turning the pages.[24] Typically also the reader sits in an enclosure, a room, a garden, or the shade of a tree.[25] The truly devoted and collected readers in fact create an aura of seclusion and concentration wherever they happen to read. Though they seem to be absorbed in their books and lost to the world, readers are actually engaged in a vigorous and consequential enterprise.

So they are at any rate when the book is good and the reader knowledgeable. There can be intensive reading in the absence of knowledge. Emotion then stands in for cognition, and where desire is not balanced by experience, where erotic longings or a lust for power have not been tempered by reality, the cheapest of novels can loosen a rush of gratification. The "reading madness" of eighteenth-century Germany, the endless devouring of amorous novels, marked the route of escape that women took whose world was confined and whose life was drained of responsibility.[26] Under such conditions, the reality conjured up by the reader resembles the virtual reality produced by information technology. The hallmark of both realms is escape and seclusion from the actual world although even then there is a difference. The reader's world is diffuse and suggestive while a virtual reality is definite and detailed.

But a text can be more than a catalyst of emotional reaction. When we read a good story intelligently, we follow the author's instruction in the construction of an imaginary world. The author gives us the blueprint, but we must supply the materials and situate the structure. The materials are our experiences as well as our aspirations. Reading Maclean's "A River Runs Through It," for example, we must draw on our experience of rivers and forests, of family, of helplessness and consolation. The location of the structure is somewhere in the life of our imagination, that realm of pregnant possibility that surrounds and informs our actual life.[27] Thus to read is to gather our past and illuminate our present. It is a focal activity that collects our world as a convex lens does and radiates back into our world as does a concave mirror. Reading at its best realizes a world view. Like a vision quest, it is solitary and outwardly passive. But in reality it vigorously engages and shapes our vision of the world. Intelligent reading of fiction and poetry, far from being an escape, is a tacit conversation with actual reality.[28]

If utilitarian reading clarifies the world and poetic reading illuminates it, then people whose literacy is low must live in a world that is confused and confined. Their comprehension of reality is restricted to what they can immediately hear and see.[29] There is no possibility of dispelling confusion by examining records, investigating dates, and perusing explanations, or by grasping the present within the context of history and comprehending facts against the foil of fiction. People who can slowly spell out words and laboriously make sense of them will be exhausted by the mechanics of word recognition long before a significant text gets to expand and clarify their world. About half of the population in this country lives in this culturally impoverished condition.[30] What is worse is that the sense of distress about this calamity has evaporated on the part of those whose literacy is crippled. Most of them believe that they can read and write well or very well.[31] Such delusion is possible only because those of us who should know better are but mildly alarmed.[32]

Realizing Information
Playing

Signification and Realization

Reading is an eminently contextual affair. Readers bring a context of intelligence to a text; the text in turn enlarges the context of the readers' reality by instructing them to place the actual world within the horizon of its past and of its commanding possibilities. Reading can do much less and much more. It can be merely entertaining and even distracting when a text does nothing but push our emotional buttons and call up those prefabricated images we call clichés. An entire industry of cheap novels is dedicated to that enterprise. Yet, reading can also be the occasion for a fundamental change of life. "Take it and read it—take it and read it," a voice said to the young Augustine (354–430), and a smart and restless student started on his way to faith and saintliness.[1] Most reading, however, lies somewhere between diversion and conversion. It tends inconspicuously to enlarge and inform the possibility space of our lives.

Traditional music involves a different way of realizing information. Music highlights the structure of signs rather than the context of things and chiefly converts time into events rather than confinement into possibility.

Music is nothing if not structure. But whether it is essentially signified structure (a score) or realized structure (a performance) is very much disputed. The dispute is fueled by the need ever and again to go back and forth between the abstraction and the realization of musical structure, to realize a score in a performance and to recognize a performance as being of a score. Not so in architecture. A building as a rule is the one-time realization of signified structure, of a design. The concrete structure obliterates its blueprint. The reading of a text, like the performance of a score, can be an ostinato affair, occurring again and again. And like a performance, reading is reversible. A poem recited can be converted back into a text or checked against it. But reading is so mundane and variable that the divide between signification and realization has become inconspicuous and fuzzy. In music, realization is still an eminent event, culturally prominent when it occurs in a concert and technically imposing when it is done in a recording studio.

The fit between abstract and realized musical structures is anything but tight. Performances of Bach cantatas vary greatly. Instead of a choir, Joshua Rifkin uses one voice for each part. Nikolaus Harnoncourt and Gustav Leonhardt have boys for the alto and soprano parts. Helmuth Rilling relies on women. His version of Cantata no. 10, for example, takes 21 minutes 30 seconds, Leonhardt's lasts 21 minutes 38 seconds, Werner Neumann sets the time of performance at 23 minutes.[2] How many of Bach's Cantata no. 10 are there? The natural reply is that there are indefinitely many performances, but they are all of one and the same cantata. And how do we identify that one cantata? By means of its score. Musical notation is a system of digital signs and therefore has the unsurpassable precision and endless durability of such a system. Nelson Goodman has highlighted these qualities of notation and stressed the supreme and untouchable status of the score—supposedly the sole warrant of the integrity of a piece of music. If we tamper with the least feature of a score, the identity of a musical work is said to be in jeopardy, "for by a series of one-note errors of omission, addition, and modification," Goodman holds, "we can go all the way from Beethoven's *Fifth Symphony* to *Three Blind Mice*."[3]

The identity and integrity of a piece of music can be underwritten by a score only if there is a complete and authoritative score. In the case of composed music, obviously the composer's original rendition should be that score. Sometimes there are calligraphically careful autographs as is true of Bach's Cantata no. 131. The original score of Cantata no. 10, however, is sketchy and incomplete. Nor was it authoritative to Bach himself. He had the duet "And mindful of his mercy" of that cantata copied for inclusion in a collection of chorales for organ. When Bach checked the printed version, he not only corrected mistakes but made changes of a purely compositional sort. As Christoph Wolff has put it, for Bach, when he "set his indefatigably self-critical hand in motion, there seemed to be no such thing as an untouchable text, whether manuscript or print."[4]

When Bach improvised on the organ or John Coltrane on the saxophone, there was no score at all, and there surely was music of the highest order. But, one might reply, there could have been a score, or someone could have made one. In the case of Coltrane, some of his recordings are in fact used as scores by aspiring jazz musicians. Whatever the vagaries of performances and variations of scores, one might conclude, there is some pure and abstract structure that underlies a piece of music, however notated or played. Reference to such a structure seems to reflect best the way we identify, remember, and discuss *the* Cantata no. 10 as distinguished from its scores and performances. If such talk, Peter Kivy has cheerfully argued, forces us to assume invisible and immutable forms or structures, eternally residing in some Platonic heaven, so be it.[5]

Bach's music seems to illustrate this view particularly well. It is a commonplace that Bach's music sounds great no matter what instruments it is played on, while Chopin's sounds awful on anything but a piano. Bach's music is pure structure whose power asserts itself regardless of instrumentation. The autograph for Cantata no. 10 comes close to reflecting just such a structure. It contains, in Robert Marshall's words, "only the 'essential structure' of a work—little more than the pitches and rhythms."[6] Similarly, the great concluding works of Bach, *The Musical Offering* and *The Art of the Fugue*, may be written for no particular instruments at all.

The notion of music as pure structure seems to be supported also by the nature of musical comprehension. There are some things about a particular piece that can be stated unequivocally and are agreed upon by all who are musically literate. Cantata no. 10 is in G minor, four-quarter time, begins with six instrumental voices that are joined after twelve measures by a seventh instrument and three singing voices and after one more measure by a fourth human voice. There is a theme, a psalm tone, carried first by the soprano voice and then by the alto.[7] The musical mind appears to command a set of formal rules—a musical grammar, so to speak—that allows it to abstract from the sounds or a score of the cantata these and other structural features.[8] Variations on a theme, for example, can be recognized only if at one level of understanding we note the recurrence of *the* theme, stripped of its particular modifications. You have to ignore some detail, set down in the score, to get the structure of a theme. In Cantata no. 10, the psalm tone that serves as a theme is first presented by the soprano and then, five steps lower and with some notes added, by the alto. It recurs as a cantus firmus, the calm and abiding melody that underlies first the intricate movement of instruments and voices in the duet and then the stately chorale of the concluding part.

Thus Cantata no. 10 can be given a structural rendition at various levels of detail. There is most detail in a performance. In fact the structure of a performance is so thick that much of it exceeds our ability to describe it unequivocally, and the profusion of it seems to conceal the underlying order of the piece. The score appears to display something like the pure structure of the music. But a summary rendition of the cantata as reflected in Bach's autograph of Cantata no. 10 will exhibit a yet purer and more abstract form, dispensing with many details of the standard score.

In the history of the West there has been a tradition where the ascent to ever purer structures was thought to reveal more and more clearly not just the nature of this or that thing, but the essence of reality entire. This essential structure was held to come closest to the surface in music, mathematics, and cosmology. Pythagoras in ancient Greece was the first to posit this order emphatically and to trace its lineaments in numerical ratios, musical harmonies, and planetary orbits. For the Pythagoreans, not only the structures of reality converged at their highest and purest levels, so did human endeavors. At the peak of Pythagorean practice, insight and pleasure, curiosity and piety, emotion and cognition were one.

In this century, Hermann Hesse has given expression to an inclusive and dramatic version of Pythagoreanism in his novel *Magister Ludi* of 1945. It portrays a future civilization whose best minds have succeeded in reducing the world's cultural creations to the formulas of a symbolic language. This elite conducts games where certain formulas are presented and worked through, much like themes in a piece of music. Hesse sketches the game again and again, from various angles and against different historical backdrops. In one such sketch he says: "Just as the pious thinkers of earlier times presented creaturely life as being on its way to God and thought of the manifold world of appearances as reaching its completion and conclusion in divine unity, so the figures and formulas of the glass bead game were building, performing and philosophizing in a cosmic language, drawn from all the sciences and arts, playing and reaching for perfection, for pure being and a completely fulfilled actuality. 'Realizing' was a favorite expression among the players, and they felt their activity to be a path from becoming to being, from the possible to the actual." [9]

A more recent and ambivalent version of Pythagorean delights has been crafted by William Gibson in *Neuromancer*, where, in an alternative future, economic and political power have been transformed into huge databases that are located in a structural manifold called *matrix* or, more famously, *cyberspace*, "bright lattices of logic unfolding across that colorless void." [10] The virtuosi of the matrix can experience cyberspace through electrodes attached to their foreheads and manipulate its structures through a keyboard computer called a *deck*. *Neuromancer*'s protagonist is such a virtuoso, a "cowboy." "He'd operated on an almost permanent adrenaline high, a byproduct of youth and proficiency, jacked into a custom cyberspace deck that projected his disembodied consciousness into the matrix." [11]

From the Pythagorean point of view, to realize the structure of signs is to disguise them at best and to corrupt them at worst. In this spirit, the purest enjoyment of music would be the one we find in the film *Amadeus* where Constanze Mozart presents Salieri with scores from her husband's hand and we see Salieri enraptured by the entirely abstract notation before him. That there is, however, a profound flaw in the Pythagorean picture of structure is suggested by the air of paradox in Hesse's allusive and Gibson's impressionistic accounts of pure order and its delights. Why paint such gauzy pictures of something that is claimed to be supremely clear and crisp?

We do have sharply etched renditions of pure structure in the language of mathematics. The purity of mathematical structures is a reflection of signs that do not refer to things but to structures simply, to lawful relations whose objects are free of empirical or causal contingency. The comprehension of such structures, as when one discovers or grasps a theorem, can certainly be a great pleasure. But this is not the structure nor the pleasure Hesse and Gibson have in mind. Mathematical structures can be applied to music or cosmology, but they do not of themselves encapsulate the essence of a cantata or the universe. As for pleasure, an organist as well as a listener will enjoy the performance of Bach's fourth Schübler chorale even if player and audience are quite familiar with the piece. I doubt many mathematicians would once more go through the proof of the Pythagorean theorem just for the delight of it. It is crucial to a piece of music that it be realized regularly and richly. There needs to be a practice of performance and audience. A piece of mathematics needs to be realized somehow and sometime, to be sure. But regularity and thickness of realization are incidental to it.

Structure and Contingency

Just as structure makes signs possible, so signs make structure visible, and it is tempting to think that if only we could devise an adequate notation, a proper system of signs, then structure as such and the structure of things would be revealed in their final purity. Awkward notation can certainly be a bad obstacle to the discovery and demonstration of all kinds of formal structures. But no claim that a definitively lucid notation has been discovered can ever be made good. Any musical score, for example, in whatever notation one wants to point to as evidence of pure structure, has answered questions in one way that could be answered in another, containing particular slurs, rests, and notes that might be different or absent. Even if one tried to triangulate a structure and were to claim that two different notations are structurally equivalent, the equivalence could never itself be presented purely and explicitly. It is as though I were to say that "3" and "III" mean the same thing, namely, *three*. But to someone who does not speak English, "three" means nothing. However you put it, *three* is embedded and realized in the contingencies of a culture and the conventions of a notation.[12]

How then do we explain that people manage to talk about the

CHAPTER NINE

same piece of music no matter how different their symbols and circumstances? As Donald Davidson has often pointed out, it is one and the same reality that asserts itself in various languages and makes sentences true or false.[13] Formal structures, musical ones in particular, must be part of that reality. It is both *one* reality and a *manifold* reality. The *how* of things unites them into one world, the *what* of things gives them their particular face and voice. Things are linked through time by causal ties and tied together at one and the same time by laws, regularities, and bonds of kinship. The laws of mathematics and physics constitute the thinnest and clearest lines of likeness and connection. As these lines get woven into the texture of more complex things, connection becomes less discernible and likeness gets coarser. Lawfulness declines to regularity, and at the highest level of human and cultural complexity reduces to mere kinship.

When we stress the how and lawfulness of a thing, we call it structure. When we intend the what and idiosyncrasy of a thing, we can call it contingency. Structure and contingency are the two principal ways reality presents itself. Yet reality is not divisible into structure and contingency without remainder. That the world is woven together from laws and data is a further contingency of lawlike necessity, and so on endlessly even though intelligibly. Reality is both knowable and unsurpassable. Positing Platonic structures is the attempt to get control of reality by dividing it all the way down, setting immutable and invisible structures to one side and the contingent remainder to the other. But any such attempt must presuppose and depend on what it tries to define and produce—signs that refer and things that are present within an inexhaustible context.

Though the ground of reality is unsurpassable, it is not incontestable and certainly not uncontested. The practitioners' agreement about musical structure is brittle and in some regards infeasible. It is the genius of musical notation that it presents music in a kind of realization that is substantial enough to capture a particular musical structure and distinguish it from all others and yet thin enough to constitute a musical agreement that transcends centuries of differences in instruments and practices and the contemporary varieties and contingencies of performance styles. But such unanimity is bought at the price of ambiguity, of attenuated significance, and at the cost of decisions delayed. In a performance, to the contrary, all the questions about instrumentation, pitch, tempo, phrasing, and more are willy-nilly answered and some inevitably in ways that cannot possibly agree from one performance to another.[14]

To perform a piece of music is to comprehend a musical structure that has emerged from the rich and definite circumstances of the composer's world and has assumed an austere and abstract reality; and it is to embody that structure again in the thick and particular setting of today. The arc that extends from composition via score to performance has a relatively fixed point at its vertex, in the commonly acknowledged score. But the lines that lead up to it from the composer's world and the lines that descend from it to conclude in our midst can be drawn in many ways. In any case, to draw the lines of musical reality into a notation is always to draw the threads of the past together and let them pass through the eye of the score so that we can tie them again into the fabric of the present.

Richard Taruskin has pointed out that performers demonstrably work in the texture of their time and culture even when, perhaps precisely when, they aim at a timelessly valid and culturally absolute performance practice.[15] Musical practice is entitled to fruitful prejudices. But theoretical reflection is better off when it comprehends its circumstances. What it generally discloses, however, is not a method that leads to the culturally appropriate performance, but an abundance of claims and invitations. We should not think of them as a closet full of costumes that we can arbitrarily pick from to drape over the austere structure of the score. Present conditions and historical particulars have an eloquence we are answerable to.

To some extent, every musical score is a reflection of its cultural environment. But kinds of music differ in the number of strands that they have gathered from their culture and entrusted to our care. Not that it is trivially obvious which of the cultural threads visible in a piece of music are intrinsic to its fabric. Bach's music has been controversial in this regard. Friedrich Blume has no use for the religious context of Bach's music and has argued that "numerous works, oratorios, masses, and cantatas, which we have grown deeply to cherish as professions of Christian faith, works on the basis of which the

Classical-Romantic tradition has taught us to revere the great churchman, the mighty Christian herald, have *a limine* nothing in common with such values and sentiments and were not written with the intention of proclaiming the composer's Christian faith, still less from a heartfelt need to do so."[16] But to this one may oppose Bach's own words: "The figured bass is the most perfect foundation of music. It is played with both hands in this way, that the left hand plays the notes prescribed, but the right hand adds consonances and dissonances so that a sweet-sounding harmony may result to the glory of God and for an allowable delight of the heart. And as in the case of all music, so also the purpose and final goal of the figured bass should be nothing else but only the glory of God and the restoration of the heart [*Recreation des Gemüths*]. Where this is not observed you have no real music, only devilish bleating and harping."[17] One need not take sides in this controversy. It is surely a matter of more or less rather than either/or. Bach's secular and instrumental music is least tied into the web of his time while his religious music, sustained and clarified by sacred texts, is most so.

Thus Cantata no. 10 was written for a special day of the year, 2 July, the feast that commemorates the Visitation of Mary. The evangelist Luke reports that Mary, pregnant with Jesus, visited her relative Elizabeth, pregnant with John who was to be the Baptist, and "when Elizabeth heard the salutation of Mary, the babe leaped in her womb."[18] Mary responded with the Magnificat, the song of every sinner, schlemiel, outcast, and underdog who has been flooded with amazing grace. For Luke who was a Gentile and at first seemed to be excluded from the blessings of the Messiah, a Jew who had been preaching to Jews, this must have come from the heart. But the Magnificat is a gesture of reverence to the Jewish tradition as well. It takes up the song first sung by Hannah in the Book of Samuel—"a psalm of national thanksgiving."[19]

Bach underscores this tie to the Hebrew past by setting the Magnificat to a psalm tone and so harks back to the music of the synagogue that extends back for millennia, and he adds to the Magnificat an elaboration of the promise first made to Abraham and retold in Luke as the story of Zacharias and his wife Elizabeth who in their old age are promised a child, John the Baptist.[20] This allusion is spelled out by Bach's unknown collaborator:

> What God of old to our forefathers
> In promise and in word did give,
> He here fulfills in all his works and deeds.
> What God to Abraham,
> When he to him into his tent did come,
> Did prophesy and promise,
> Was, when the time had been fulfilled, accomplished.
> His seed in truth must be as far
> As ocean sands
> And starry firmament extended;
> For born was then the Savior,
> Eternal word was seen in flesh appearing,
> That thus the human race from death and ev'ry evil
> And also Satan's slavery
> Through purest love might be delivered;
> So it remains:
> The word of God is full of grace and truth.[21]

This appropriation of Abraham's story, together with the score, is an engaging demonstration of what it means to bring a text to life.[22] The first part is sung by the tenor as a *recitativo secco*, the dry if stately sing-song, austerely accompanied by organ or harpsichord and commonly used to convey a narrative. But when the story reaches the point that is so vital to Bach, "when the time had been fulfilled," violin and viola voices join in to celebrate the fulfillment of the promise with lively and festive figures, and the recitative rises to something like an aria.

Taking up the religious ties in Bach's cantatas does not accord well with the modern temper, particularly when a cantata contemplates, if it does not celebrate, the hopelessness of this vale of tears and the helplessness of human reason to cope with it. Richard Taruskin has called attention to this challenge and expressed his admiration of Gustav Leonhardt and Nikolaus Harnoncourt "for their refusal to flinch in the face of Bach's contempt for the world and all its creatures."[23] Taruskin had earlier rejected the eighteenth-century notion that music is nothing but "the art of pleasing by the succession and combination of agreeable sounds."[24] What Taruskin opposes is not the claim that music is sound, but that it must please and gratify. But as long as music is just sound, does it matter much whether it sounds pleasing and agreeable as the eighteenth century has it or "loathsome and disgraceful" as Taruskin insists on behalf of certain Bach cantatas? Is music not also something that is done at a particular place and time in a particular community—a place, for example, consecrated to worship, a time designated for celebration, a community that affirms the caducity of this world and the glory of another? There may be no such circumstances today. Performances that rise to the challenge of Bach's scores, Taruskin holds, "could never work in the concert hall, it goes without saying, and who has time for church?" So what is the answer? "But that is why there are records," Taruskin concludes.[25]

Records, however, submerge the full structure of information more resolutely even than writing and occlude the place, the time, the ardor, and the grandeur that provide the setting for the musical realization of structure. For a score to become real, it requires not only its proper place and time but also a communal tradition of extraordinary discipline and training. Human beings need to struggle with the recalcitrance of things and the awkwardness of their bodies before the ease and grace of music making descend upon them. The physical reality of a flute enforces stringent demands on the posture, the breathing, and the embouchure of the student. The violin requires that players place their fingertips on the uncharted territory of the finger board with a precision of fractions of a millimeter. Ordinary mortals need forceful teachers and many years of practice to learn these skills.

Once they are trained, however, musicians give voice to the grandeur of reality. They bring out the common and concealed kinship of movement in things and raise the resonance of reality to singing. When you see swallows circle and dive to the strains of Mozart's clarinet concerto, you notice a coincidence of power and grace, a consonance of gliding and turning, a serene suspension of wings and sounds above the troubled world. But music need not be so ethereal. It has many ties to reality. It can move like a dancer, rise like a fortress, and cry like a baby.

Both the ardor and the grandeur of music give structure to time. Practicing and performing require regularity and engender focal practices. At its most eminent, however, music rises from its practices to an event that constitutes a landmark in time. It can be a spectacular event as when a star sings before hundreds of thousands in Central Park. But the occasion can be less worldly and more solemn at once—a congregation of some 300 rather than a crowd of 500,000, a brave community ensemble instead of a celebrated tenor, the piety of Bach in place of Puccini's pyrotechnics. Toward the end of the liturgical and calendar year of 1995, the choir of St. Paul's and a group of local players and musicians in Missoula, Montana, performed Cantata no. 140, which begins with the hymn, "Sleepers, Wake." And needless to say, there is kindred music every week all over the country in synagogues, temples, and churches, ranging from the venerable cantillations of the Hebrew scriptures to the glories of gospel music.

Realizing Information
Building

Construction and Contingency

In building, we realize information most tangibly and publicly. Here information is truly an instruction for the construction of something real, and what has been realized guides and constrains the daily round of life more obviously than reading and more unyieldingly than music. But precisely because building engages reality so vigorously, it is also most at the mercy of contingency and most revealing about it.

While structure provides the drainage for the flow of information, contingency is its wellspring. If the universe entire had a crystalline structure, there would be little to find out and report about it. But as it is, reality addresses and sometimes assaults us in unpredictable ways. Our sense for the force of reality has hardened, however. We tend to think of reality chiefly as material that is ours to shape. Contemporary thought, in particular, has little regard for the expressions of reality. Still, contingency is the one concession thoughtful theorists make to the eloquence of the world. Philosophers like Daniel Dennett and Richard Rorty and scientists like Stephen Jay Gould and Steven Weinberg recognize that the fabric of reality is not to be explained by laws alone but through data, givens; and they acknowledge that the irregular residue of reality is more than inert or featureless stuff, that it bucks up against our plans and predictions.[1] Yet the tendency of mainstream thought is to reduce the component of givenness and sheer presence to randomness and meaninglessness.[2] Contingency, however, is inherently meaningful and so makes significant information possible. Contingency comes to us as misfortune or good luck, as disaster or relief, as misery or grace. Only when contingency is artificially confined or refined is there something like strict randomness.

Building, more so than reading or playing, runs into the perils and favors of contingency. Any attempt to show this generally, however, would be self-defeating. It would miss the contingency of contingency, the presence of the unforethinkable. But things have reference as well as presence, and for a particular case to be informative it needs to have not only the gravity of the particular but also reference to more general conditions, ours included.

Perhaps the construction of a medieval church that has endured to this day can have that sort of presence and reference. So consider the building of Freiburg Minster in the Upper Rhine Valley. Taken as a whole, the church appears to strike a perfect balance of energy and harmony. The great tower at the west end of the church rises to a height of 372 feet, exactly equaling the length of the church, and it would overpower the main part of the church, the nave, were it not counterbalanced by the concluding eastern part, the choir, a little taller than the nave and equal to it in length. Where the choir connects to the nave, it is flanked by two smaller towers that reflect in their construction the great single tower over the western facade. Looking at the minster from the south, one sees the huge tower at the left, followed at one third of the tower's height by the nave, the end of the nave being marked by the small towers rising to half the height of the west tower, the roof of the choir emerging between the two towers, extending to the right and sloping down to conclude the church (see fig. 11).[3] Victor Hugo was moved to compare it to the more worldly Strasbourg Minster lower on the Rhine: "It has, with a different design, the same elegance, the same boldness, the same verve, the same mass of russet and somber stone, pierced here and there by luminous openings of every shape and size."[4]

It would be natural to assume that such power and coherence are the result of an inspired and comprehensive design. Designs, after all, are the means humans employ to overcome the contingencies of building materials and practices and to impose their will on the vagaries of the world. As it happened, the church was begun at the turn from the twelfth to the thirteenth century when intellectual and architectural design had risen to a new level of explicitness and so-

FIGURE 11   The south side of Freiburg Minster after a nineteenth-century engraving

phistication. Both in theology and architecture, the practice of piling up materials in sturdy and customary arrangements gave way to thinking things through and articulating them in a perspicuous and elegant design. In theology this line of development goes from the compilation of biblical sentences under thematic headings to the carefully and grandly constructed *summa* where every word has its systematic meaning and place. In architecture the trajectory extends from the massive and ponderous Romanesque to the highly articulated and soaring Gothic.[5]

Twelve hundred is the midpoint of the one hundred years that comprised this intellectual quickening. Its geographical center was Paris. The east side of the Upper Rhine Valley, some two hundred miles to the southeast of Paris, was a cultural backwater. When around 1200 the duke of Zähringen and the citizens of Freiburg began to build a new parish church, its design was in the Romanesque style although the pioneering monuments of the Gothic style had already been erected in their essentials. The distinctive and innovative parts of Saint Denis and all of Notre Dame save the roof had been finished by then. Communication among the master masons, moreover, was rapid and intensive.

What happened to the initial design can be gathered from the building itself, provided it is placed in the context of historical information. The minster, in fact, is not only a focal point of the information that has been realized in it, but also of the information that has been collected about it, and it illustrates not only how building realizes information but also how reading realizes the space a building needs to become intelligible.[6]

At the start of the minster's construction, around 1200, at any rate, there must have been artful if somewhat provincial and old-fashioned drawings for the construction of a Romanesque church. Few plans of medieval churches prior to the fourteenth century have survived. No originals exist of Freiburg Minster though there are copies of the tower design that vary the original for local purposes.[7] We know from the grand design for the monastery of St. Gall that superb and highly informative plans existed by the early ninth century, and we know from the sketchbook of Villard de Honnecourt of the first half of the thirteenth century that medieval master masons were accomplished and inveterate practitioners of drawing and design.[8] We do have a fair idea of the original design of the minster from carefully reconstructed models, ground plans, elevations, and sections of the eastern part of the church.[9]

That is where construction began, more particularly at the place where the nave was to intersect with the shorter transverse section, the transept, that, together with nave and choir, gives a church its characteristic cruciform floor plan. The intersection, called the crossing, was finished and surmounted by an octagonal dome. The transept and a fairly shallow choir were completed as well. In the corners where choir and transept meet, the lower stories of two towers were erected.

All this took two or three decades, and then the news of the Gothic style swept away the Romanesque design. Awkward attempts were made to continue the church toward the west facade in the bolder and loftier style. Around 1240 a master mason, truly skilled in the Gothic style, arrived on the scene, completed most of the nave, and began the single western tower, a nearly solid block on a square ground plan. Around 1280 a new master mason took over. Most likely he came from Strasbourg Minster where Gothic construction had achieved a high degree of sophistication and in time produced the steeple that Thomas Jefferson in 1788 found to be "the highest in the world, and the handsomest."[10] It was this master who changed his predecessor's upper design of the tower from a massive and measured construction to an intricate and lofty sculpture.

Around 1340, the smaller towers that had been built in the Romanesque style and risen to half of the roof's height were each given an additional Gothic story and a spire of stone tracery like that of the great tower. In the year 1354 ground was broken for the late Gothic choir to replace the Romanesque apse. Around 1370, construction stopped for a century to be resumed in 1471 and completed more or less in 1513 when the new choir was dedicated.

From the standpoint of rigorous and total design, the construction of the church is a history of false starts and aborted plans, of incongruities and accidents, of contingency in the sense of *coincidence*.[11] But this point of view fails to reveal how the several parts and periods of construction came to converge in a coherent whole.

In an older sense of the word, contingency signifies the way things make contact with one another and coalesce. Victor Hugo had a keen sense of how contingency as *convergence* has shaped Freiburg Minster: "The whole lower part of the church is Romanesque as are the two lateral portals, one of which, the one on the right, is masked by a portico from the Renaissance. There is nothing more remarkable, as far as I am concerned, than the encounter of the Romanesque style and the Renaissance style. The Byzantine archivolt, that strikes us as so austere, and the Neo-Roman archivolt, that looks so elegant, collide and connect with one another, and that lower structure embraces these fanciful creatures with harmony and sees to it that they touch each other without injuring one another." [12]

Contingency, finally, has a still older meaning, or so does the verb *contingere* that contingency is derived from. In the sense in which Cicero occasionally used the verb, contingency means something like *consummation*, the happy completion of a task; and much later, in a Latin translation of Matthew's gospel, contingency is the divine grace that lends eternal validity to human conventions. "Again I say unto you," the passage says, "[t]hat if two of you shall agree on earth as touching any thing that they shall ask, it shall be done [*continget*] for them of my Father which is in heaven." [13]

High contingency is principally embodied in the great tower of the church. It is a consummate work of art as well as a civic and spiritual landmark. Architecturally considered, it begins in its lower third as a nearly solid base on a square footing, supported by largely unadorned buttresses, with the grand portal and a few windows carefully inscribed in the plain walls (see fig. 12).[14] The middle third is part of the new design, an octagon that is finely articulated and, in its upper half, consists of an open structure of eight tall, pointed arches. The final third is the spire, composed of eight slender, isosceles triangles of delicate stone tracery. At the seams between these three parts, the design both disguises and prepares the transition from one stage to the next. The base and the octagon meet below a twelve-cornered gallery that marks both the four corners of the base and the eight of the octagon. The lower half of the octagon is flanked at the four corners of the base by triangular piers that terminate in tabernacles and spires to reveal the octagon proper that in turn ex-

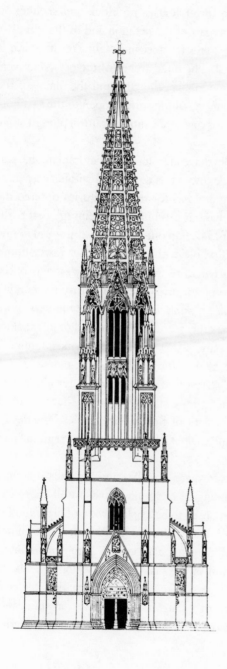

FIGURE 12   The west elevation of Freiburg Minster, from Pius Enderle, *Zahl, Klang, Licht* (Frieburg: Schillinger, 1993)

tends above the base of the spire through spurs on the eight corners and gables over the great arches. Thus, the parts emerge from each other in a sequence from solid density to intricate perspicuity.

The upper half of the tower houses no clock and shelters no bells. It consists of the open part of the octagon and of the spire—a vertiginous inner space of some 180 feet in height. Yet its space is external as much as internal. The huge arches of the octagon reveal a panoramic view of the mountains, the city, and the plains of the Upper Rhine Valley. The filigree of the spire is open to the sky. Formally considered, the tower is one of the most beautiful piles of stone put up by human hands.

Design and contingency have amounted to a moral landmark as well as to a work of art. The tower was completed within two centuries of Freiburg's founding and has for many centuries ordered the works and days of its citizens. It is prominent in Freiburg's first faithful portrayal, a woodcut of 1549, and one will be hard-pressed to find a picture of the town as a whole without the minster tower dominating the scene. The church entire marks the east-west axis of the compass, exhibiting the common orientation of Christian churches "toward the rising sun," *ad solem orientem*—whence the very word "orientation." In the minster this feature is particularly radiant since the choir, the easternmost part, is also the loftiest and most luminous region of the inner space of the church. The tower moreover is a focal point of the region. A single western tower, after all, is very unlike the French cathedrals whose stately two-tower-facades recall and heighten the imperial might of Roman city gates.[15] The design of the minster monumentalizes the simpler structure of regional parish churches.[16]

As the tower has ordered space, so it has time. Its bells announce the time of day, the beginning of daily mass, and in full chorus, of high mass on Sunday. There used to be a bell to warn of fires, another to indicate baptism, and yet another for funerals. In fact, the church itself, in its longitudinal dimension, can be seen as the embodiment of time and the progression of the history of salvation.[17] Thus the minster and its tower work much like a natural landmark—the peak of a mountain or the portal of a canyon—and convey an abundance of inconspicuous and incidental information. Their very material,

CHAPTER TEN

the reddish-brown sandstone, provides information about the local geology, about the forces that ground rocks of silicon dioxide into sand, mixed it with iron oxide, and compacted it back into stone.

But unlike a monument of nature and much like a book, the minster is a thing containing signs. The medieval cathedral, as Victor Hugo put it, is "the book of stone."[18] And to be sure, to a well-instructed Christian and to anyone familiar with the Hebrew and Christian traditions, the minster is an open book. The powerfully welcoming portal at the base of the tower leads into a vestibule lined with the many figures that mark the history of salvation, among them Abraham and Isaac who are represented twice, once among the great figures on the left wall as one enters and once in the upper right arch over the doors that lead to the interior of the church. The outermost jamb of that arch is adorned with the statues of a buoyant *Ecclesia* on the left and a downcast *Synagoga* on the right, Church and Synagogue, the Christian and the Hebrew traditions.[19] Entering and progressing toward the choir, one is accompanied by the apostles attached to the piers of the nave. The focal point of the choir is occupied by the high altar and by its painting of the coronation of Our Lady after whom the church is named Our Lady's Minster. The walls of the choir are hung with five great seventeenth-century tapestries portraying the story of Abraham and Sarah. Beside its cosmic and historic messages, the local versions of foot, yard, and bushel were chiseled into the western buttresses of the tower as were the outlines of loaves, tiles, and a brick.

The Ambiguities of Signs and Things

No design can specify its realization fully. To convey exactly as much information as the thing realized, a design would have to exhibit just as many features as the thing. But then it would be a duplicate of, rather than information for, a thing. All cultural signs, or symbols as we may call them, underdetermine their realization and are in this sense necessarily ambiguous. This kind of symbolic ambiguity—an ambiguity of austerity—makes for the creativity of realization and the drama of contingency. When the creative power of humans and the contingency of reality are consummated in a great work, the latter has a presence more commanding and expressive than that of a text, a score, or a plan.

But what exactly does a significant thing say? Alas, it says many things at once, and humans are not always sensible and equal to its crucial meaning. Thus the eloquence of a thing is easily mistaken. To the austerity of symbolic ambiguity there corresponds the mirror image of real ambiguity, the profusion of meaning that a powerful thing radiates. Things have the richness and particularity of a picture while conventional signs have the precision and generality of a concept.

When meaning began to decline early in the modern era, the profuse ambiguity of natural things and works of art came to compare poorly with the austere definition of printed information. If Victor Hugo is right, the latter in fact hastened the demise of the former. Late in the fifteenth century, when the last parts of the minster were still being completed, the archdeacon of Notre Dame in Hugo's novel "gazed at the gigantic edifice for some time in silence, then extending his right hand, with a sigh, towards the printed book which lay open on the table, and his left towards Notre-Dame, and turning a sad glance from the book to the church, 'Alas,' he said, 'this will kill that.'" [20]

In any event, the shift of cultural energy from the presence of things to the reference of signs, from meaning to information, was part of the large contingency of historical change. A characteristic eddy in this broad stream once more revolves about the tower of Freiburg Minster. From the early nineteenth century on, a broad segment of art historians was unable to acknowledge the high contingency of the tower, the convergence of accidental factors in a spiritual landmark. Beauty, to these theorists, was no longer conceivable as commanding presence. In its stead they searched for the secret of the tower's underlying formula that would assign each part its relation to every other part. It was the attempt to reveal necessity beneath complexity and reference beneath presence.

To be sure, there was enough inspiration from classical proportionalism and enough mathematical symbolism in medieval Christianity to set one on the wrong track. As for numbers there is the *one* true God, the *three* persons of the Trinity, the *four* gospels, the *seven* sacraments, the *twelve* apostles, and more. These numbers could be connected to the Pythagorean ratios that had originated in Vedic altar construction and come down to the medievals via Plato and Augustine: 1:1, 1:2, 2:3, 3:4, 4:5. The ratios in turn can be translated into plane and solid figures. Moreover, the proportional specifications of the Salomonic Temple and the heavenly Jerusalem in sacred scripture encouraged and seemed to sanction this procedure. Even God the creator was depicted in a medieval Bible as designing the world with a compass according to "true measure." [21]

The notion that medieval architecture rests on a system of proportions arose early in the nineteenth century with the romantic rediscovery and appreciation of the Gothic style. At the same time, though this is never said explicitly and was likely done unknowingly, the champions of proportional design tried to undo what had been brought to its grand solution by Descartes, the interplay and combination of magnitude and number, of geometry and arithmetic, of construction and computation. The "proportionalists" claimed that medieval and especially Gothic plans were entirely drawn as a system of proportions. Thus the nave and crossing of a church might be a sequence of squares, the transepts each half a square, the aisles a sequence of quarter squares, the choir one large square plus three sides of an octagon, and so on to the last and least detail of every arch, column, and window. Such a design would not require the employment of cardinal units of length, such as inches, and it likewise could be executed on the ground without measures such as feet and yards. All that had to be done was to mark off the length of the first great square. The rest was a matter of stakes and ropes.

In this view, the Gothic builders showed a sublime indifference and grand superiority to the kind of counting and calculating that the romantics deplored in the rational and industrial temper of their times. The tower of Freiburg Minster was the object of the most devoted and ingenious attempts at deciphering the system of proportions that contained the secret of its beauty. At least twelve different proposals have been made, ranging from a straightforward employment of the golden section to an elaborate grid of equilateral triangles. Twelve incompatible answers alone might lead one to doubt whether the principal question was well-conceived. But the historical evidence is much more damaging. The Gothic builders in fact drew plans to scale and used yardsticks to execute them.[22]

It is true, however, that they delighted in geometrical play and thought of themselves as geometricians. Villard de Honnecourt in his sketch book used geometrical grids and proportions to structure not only buildings but also faces, bodies, and configurations of persons.[23] It is plain that straightedge and compass were used to outline the structure of the minster tower on parchment. And though the originals have been lost, today's renditions of ground plans, elevations, and sections reveal the exquisite formal refinement and beauty of the design. Proportions and regular figures abound, particularly in the design of the choir (see fig. 13). But the drawings of a master mason were in part unencumbered play with geometrical forms and in part compromises with the contingencies of accomplished facts, irreversible errors, and economic limits.[24]

Real ambiguity has a more difficult and distressing aspect yet than appears in the misguided ruminations of art historians. It is part of the darkest side of contingency that the spiritual force of Freiburg Minster was in a crucial regard of no help even during the Middle Ages when the tower, in the words of the poet Reinhold Schneider (1903–58), was a "great prayer of a faithful time."[25]

The citizens of Freiburg greatly honored Abraham in the furnishings of the minster, and the portrayals of the Hebrew tradition in the minster generally reflect the genial and cooperative relations that obtained between Jewish and Christian inhabitants of Freiburg roughly from the middle of the thirteenth to the middle of the fourteenth century.[26] But from then on the children of Abraham were relentlessly persecuted and harassed for half a millennium. Like other cities in the Upper Rhine Valley, Freiburg accused its Jewish citizens of having caused or aggravated the black death of 1347–49. Confessions were pressed from them by torture. The adults were herded into houses and the houses set on fire.[27]

But beyond these common atrocities, the town of Freiburg distinguished itself by making residence for Jewish people impossible and temporary stays cumbersome. Freiburg ignored and defied orders of tolerance and freedom issued by royal and feudal edicts. It was not until 1862 that state law forced Freiburg to grant full civil

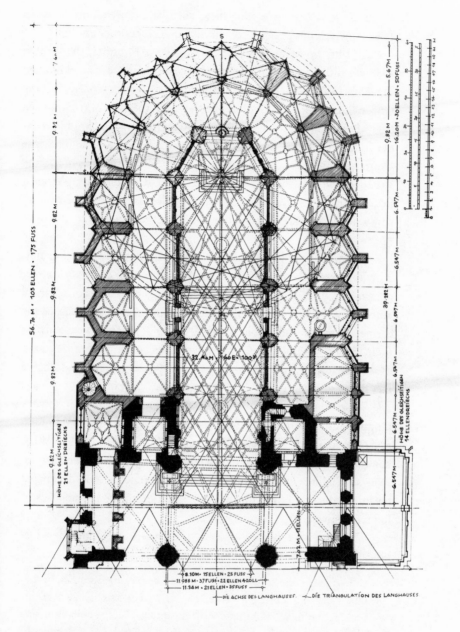

FIGURE 13   The plan of the choir of Freiburg Minster with the geometrical relations indicated, from Pius Enderle, *Zahl, Klang, Licht* (Frieburg: Schillinger, 1993)

rights to Jewish people. But then, surprisingly, relations between Jews and Gentiles became cooperative and sometimes cordial.[28]

All this, however, went up in flames on 10 and 11 November 1938 when the Nazis looted and destroyed Jewish stores and torched the synagogue that Freiburg had built for its Jewish citizens.[29] If the citizens of Freiburg did not cheer, they did not intervene either. At best, there was despondent paralysis, captured by Reinhold Schneider in his memoirs: "On the day of the attack on the synagogue, the Church should have stepped forward as the sister of the Synagogue. It is decisive that this did not happen. But what did I do? When I heard of the fires, the looting, the atrocities, I locked myself in my study, too cowardly to confront what had happened and to speak up."[30]

The ensuing holocaust has been the darkest issue of contingency. The horrors of the persecutions and concentration camps are surely contingent. It is important and illuminating to trace the historical antecedents of this catastrophe. But the latter inevitably exceeds the former in terror and significance. In history, effects always surpass their causes. The holocaust, moreover, was contingent in the sense of the convergence and consummation of evil. It was not just a brute fact but significant in the most terrible way.

When the war was approaching its end, the minster tower's moral fragility came to be matched by physical fragility. Bombing raids had to be expected, and the minster might well be leveled. What could possibly be the meaning of all this? Again the tower can be seen as a telling instance of a broad development—the decline of meaning, the rise of disconsolate ambiguity, and the possibilities of human testimony.

Charles Taylor has both stressed the loss of encompassing significance and held out the possibility of solace:

> Virtually nothing in the domain of mythology, metaphysics, or theology stands in this fashion [i.e., of earlier ages] as publicly available background today. But that doesn't mean that there is nothing in any of those domains that poets may not want to reach out to in order to say what they want to say, no moral sources they descry there that they want to open for us. What it does mean is that their opening these domains, in default of being a move against a firm background, is an

articulation of personal vision. It is one that we might come to partake in as well, as a personal vision; but it can never become again an invoking of public references, short of an almost unimaginable return—some might say 'regression'—to a new age of faith.[31]

To put Taylor's view in our perspective, reality has fallen silent, but poets can still be its advocates. To get beyond the dissolution of real ambiguity we must reach for the definition and allow for the austerity of symbolic ambiguity.

What then, we may ask, could have been the meaning of the minster tower to someone like the poet Schneider in the midst of the war and the holocaust? It must have seemed a luminous trace in desolate times, like the rays of the setting sun that are caught by the tower when the city is already dark, an inspiration, perhaps, and a sign of hope that the terrors of the war might reveal to people the truth of their transgressions. But would the tower itself be spared from destruction?

Schneider was not allowed to publish during the war. His sonnets of grief and solace were mimeographed on miserable paper, bound together—no place, no date—and circulated in the war zones and at home—samizdats before their time.[32] One such sonnet, written in January of 1944, was this:

To the Tower of Freiburg Minster

Stand unshaken, grand in spirit,
Great prayer of a faithful time.
How the day's grandeur glorifies you
When long the grandeur of the day has died.

So I will pray that I may guard faithfully
The holiness you cast into the strife.
And I will be a tower in the darkness,
A bearer of the light that flowered for the world.

And should I fall in this great storm,
Then be it for the sake of towers soaring,
And that my people may become the torch of truth.

You shall not fall, beloved tower,
Yet should the judge's lightning shatter you,
Rise still more boldly in our prayers from the earth.[33]

Ten months later, on 27 November, within twenty-five minutes, 1723 tons of explosives rained down on Freiburg. The inner city was leveled; nearly three thousand people died, and over nine thousand were injured.[34] Yet, the tower of the minster remained standing in the rubble.

Why was the tower saved? At first it was believed that the Royal Air Force had marked the cathedral with flares and carefully avoided it while hitting almost all the adjacent buildings. But later research revealed that the Royal Air Force was unconcerned and would have been unable to execute such precise bombing.[35] There was no reason why the minster was spared. It was the sheerest accident. And yet the citizens of Freiburg drew strength from the survival of the minster as they endured the end of the war and the hunger and cold of the first years after.

Today the physical perils that beset the church are typically subtle and insidious. Sulfur dioxide from air pollution mixes with rain water and drains into the pores of the sand stone. It dissolves the glue that holds the grains of sand together and erodes the features of statues and the solidity of walls and buttresses. As stones are replaced, the minster is more and more becoming a copy of itself. The minster's spiritual identity too is beset by ambiguity. At times it seems more like a tourist attraction and a hall for organ concerts than a house of worship.

When Richard Taruskin saw that the communal context for the realization of Bach's cantatas had evaporated, he suggested that we entrust the cantatas to the hyperreal perfection of compact discs and give them the imperishable existence of an insect in amber. Similarly, while the material and spiritual substance of the minster tower are coming into question, its structural identity has been captured with total transparency and control. In April of 1995, a computerized camera, mounted on a helicopter, took 1400 pictures of the tower that capture it in details down to a millimeter and in three dimensions. "In case of a catastrophe or the destruction of parts of the building," we are assured, "a precise reconstruction would be possible by means of these documents."[36]

Still, there are occasions of high contingency, moments of celebration and, if there is such a thing, of redemption, when humanity and reality seem gracefully joined. So they were on Pentecost 1997 when some ten thousand faithful gathered in the church, and orchestra and choir performed Mozart's *Credo Mass*, the kind of music that makes pillars soar and arches vault.

Part Three

Technological Information
Information as Reality

Elementary Measures

The Electron

Plato was among those early philosophers who tried to subordinate contingency to structure. The leading idea was to reduce what strikes us as marvelous or mysterious to an underlying simplicity and lawfulness. In the *Timaeus,* Plato tried to build up the world of direct experience from the regular solids that in turn, he thought, were constructed from two kinds of triangles.[1] The things and processes of the visible world he explained as compounds and transformations of the elementary particles. What seems immediate and ultimate to the untutored is said to be reducible to the workings of the elements and their components. Thus "as regards all flowings of waters, and fallings of thunderbolts, and the marvels concerning the attraction of amber and of the Heraclean stone—not one of all these ever possesses any real power of attraction; but the fact that there is no void, and that these bodies propel themselves round one into another, and that according as they separate or unite they all exchange places and proceed severally each to its own region,—it is by means of these complex and reciprocal processes that such marvels are wrought, as will be evident to him who investigates them properly."[2]

The Greek word for amber is *electron,* and the Heraclean stone is a magnet. The "complex and reciprocal processes" that underlie electricity and magnetism remained concealed for two and half millennia, and the kind of scientist "who investigates them properly" did not appear until the beginning of the modern period.[3] The first to give a sober and comprehensive account of what was known about the attractive force of amber and magnets was William Gilbert (1544–1603), physician to Queen Elizabeth I. It was he who coined the term "electric" to designate the attractive force of amber. Conjectures, experiments, and observations yielded an ever more systematic theory of electricity. The first to see a precise link between electricity and magnetism was the Danish physicist Hans Christian Oersted (1777–1851).

One eddy in this stream of experimentation was the strange glow that Francis Hauksbee (1666–1713) observed in partially evacuated glass jars that were attached to a source of electricity. Later experimenters came to realize that something like rays emanated from the source of negative electricity, the cathode, when it was placed within a nearly evacuated glass vessel, and that the rays left a glowing point on the glass near the cathode. In 1897, Joseph John Thomson (1856–1940) let a bundle of cathode rays pass between two metal plates and, by applying electric and magnetic forces through the plates to the cathode rays, caused the rays to be deflected so that the glowing spot shifted its position at the end of the glass tube.

In time, engineers transformed the end of the tube into a large, fluorescent surface. They learned how to direct the cathode ray back and forth, tracing line after line on the surface, now called the screen; and they varied the strength of the ray to generate a pattern of light and dark points. And so the display device for television and computers was born.

Thomson, however, was interested in science rather than technology. He was curious about the underlying structure of things, and not simply about the order of that structure but about its measures. Specifically he wanted to determine the quantitative structure of the cathode rays. The obvious framework for an analysis of phenomena involving forces, such as the electric and magnetic ones, was Newton's second law of motion. It tells us that the force, F, needed to produce acceleration, a, in an object of a certain mass, m, is proportional to the product of that mass and acceleration:[4]

$$F = m \cdot a$$

Acceleration, in turn, is the ratio of change of velocity, v, of an accelerated object to the time, t, that has elapsed during the acceleration:

$$a = dv / dt$$

CHAPTER ELEVEN

Velocity, finally, is the ratio of the distance, s, some object has traveled to the time, t, that has elapsed during the travel:

$v = ds \, / \, dt$

Thus Newtonian physics provided Thomson with a structure that lawfully related time and space, motion and mass, acceleration and force to one another and allowed him to take and make sense of his measurements. What Thomson was able to determine or measure was the strength of the magnetic and electric forces, the distance between the location of the plates that deflected the cathode rays and the end of the tube where the glowing spot appeared, and the amount of displacement of the spot on the end of the tube, caused by the deflection of the ray.

What Thomson finally found was not some particle of a certain mass and a definite charge, but only one number that expressed a ratio of mass to charge. Thomson, however, leapt to the conclusion, rightly if boldly as we know now, that the ratio informed us about a particle of the atom. The discovery showed that the atom was not at the bottom of the structure of reality but is composed of smaller particles. Though the electron, as it came to be called, was the first of the atomic particles, it, unlike the proton, for example, has remained elementary to this day.[5]

As it turned out, the electron is one member of a family of truly elementary atomic particles, the leptons. The other family, the hadrons, are composites of the truly elementary quarks. What appears irreducible today may tomorrow be revealed as a composite of yet smaller particles such as the so-called strings or membranes—one set of lawfully related objects being reduced to one that is still more elementary. In any event, it will be structure all the way down. No one expects to discover continuous measureless stuff at the bottom of things.

The prominence and precision that the notions of structure, measure, and reduction to elementary particles had attained in atomic physics in Thomson's time must have left a deep impression on the common understanding of information. Equally inspiring was the preservation of the significance of laws and objects regardless of mathematical manipulation. Thomson could substitute distance over velocity for time without having to worry about the significance of time disappearing from his equation. He did not have to ask himself what it really meant when a velocity squared in the denominator was reduced to a simple velocity through a substitution that inserted velocity in the numerator.[6] Algebra in this case is the perfect semantic machine where, if you take care of the mechanics, meaning takes care of itself.[7] Thus physics and mathematics engendered the hope that, once the elementary particles of information had been found, information theory might devise a semantic engine that would bridge the gap between structural and instructive information.

The electron, like the letter, turned out to be not only a vessel for the storage and transmission but also a model for the analysis and synthesis of information. But if information is a ward of the electron, the converse is true as well. Information is the guardian of the electron. No electron has ever presented itself in person. We only know electrons from the information we have about them.

The Bit

Taking the electron as its model, the analysis of information is concerned with the discovery of the elementary particles or units of information. But if information is a relation rather than a thing, it is far from obvious what in information is reducible or measurable and what is not. The most definite term of the information relation is the sign, and here reduction, if not measurement, is a feasible enterprise. Among signs, in turn, it is the system of conventional signs we call writing that lends itself best to analysis.

The first signs of writing were logographs, and logographic writing developed into large and complex systems. However, to have thousands of indivisible and unresolvable logographs is, it seems, to miss something crucial about systems of signs and vehicles of information. To resolve these thousands of symbols into a few hundred syllables is to shed some light on the underlying simplicity. To go from hundreds of syllabic signs to some two dozen letters is to obtain more clarity yet. But is twenty-six the least number of signs needed to record and convey information?

One might think that the sound structure of language requires

26 or so symbols. Yet letters are not snapshots of the infinite variety in which people pronounce and intonate words, but the result of a systematic simplification, balanced between fidelity to the acoustic reality of speech and parsimony for the sake of efficiency. In ancient Greece, the balance came to rest on 24 letters. A phonetic rendering of English would require some 40 symbols. The American Standard Code for Information Interchange (ASCII) contains 82 symbols, 26 lower and 26 upper case letters, 10 number signs for the decimal symbols, and 20 punctuation and function signs. We could do with 26 signs if we did without punctuation and lower case letters (as the first alphabetic writers did) and rendered function and number signs in letters, + as PLUS, 12 as TWELVE, and so on. But is this the limit of notational economy?

As it happened, the path to the least and fewest signs followed the course of numeracy rather than literacy. Here the question is how many number signs are needed. Counting things in groups of two seems to have been the original and originally most common way of keeping track of the how-many. Evidently two signs should suffice. However, of the competing systems of counting, in twos, tens, twelves, twenty-sixes, and so forth, the decimal system with its ten number signs prevailed and became so natural to us that for the ordinary numerate person it is hard to understand systems of a base other than ten and harder yet to get an intuitive feel for the way they work. When Leibniz (1646–1716) in 1696 or 1697 discovered the binary system of two number signs—or, more correctly, rediscovered it for the modern period—he was so taken with his finding that he petitioned Rudolph August, Duke of Brunswick, to have a medal struck in commemoration of the discovery.[8]

Leibniz was right on one crucial point. The binary system is unique among all others in constituting an irreducible system of the least signs. It is commonly used as a counting and computing system of two number signs, 0 and 1. When used in this way, the rightmost place in a number is read as it is in the decimal system. But the second place to the left counts groups of twos rather than tens, the third place groups of four, the fourth place groups of eight, and so on. Thus in binary notation, going from left to right, 11 is one group of two and one group of one equaling three and not, as in the decimal system, one group of ten and one group of one equaling eleven. Eleven in binary would be 1011, that is, one group of eight, zero groups of four, one group of two and one group of one equaling eleven. Letters, decimal number signs, punctuation and function signs can be rendered in binary notation by assigning a specific number to each character, and that is what the American Standard Code for Information Interchange does. In fact, a binary system of signs is sufficient to express anything that can be rendered in any notation whatever.

But are two signs necessary to record and manipulate information? If one kind of element, the brick, is sufficient to build all kinds of structures, why should not one kind of symbol—a notch on a tally bone or a slash mark on paper—suffice to set down and process all kinds of information? In the abstract this is possible. We could keep count of the sheep in a herd through notches on a stick, add the number of sheep in one herd to that of another by laying two tallies end to end; we could multiply by placing two tallies at a right angle one to the other, filling in tallies to convert the angle into an array, and recording the resulting total of notches on one large stick, and so on. However, if we wanted to talk about the numbers we were working with, we would need as many unique names as there would be numbers, and we would have to memorize them in sequence. Evidently, all of us have succeeded in doing this up to twenty-six or however many different signs in the case of the alphabet, and we could surely memorize much longer strings. But now we are really going back on the initial assumption since we have surreptitiously introduced twenty-six signs or more to cope with the featureless things of notches or slashes.

The human ability to identify and remember strings of uniform symbols is amazingly limited. People can accurately and at a glance identify strings of identical marks that number up to seven. Above seven they begin to estimate the number and do so less and less precisely as the number increases.[9] To grasp quantities, we need to shape and order them, put them in groups, the groups into groups, and so on. When we mark slashes on paper to count something, we usually make four vertical ones, make the fifth horizontally through the four (to stay below the treacherous seven), and connect our groups of five with the familiar grouping system of ten—the decimal system. Hence if we want both the perspicuity of a grouping system *and* the minimum of symbols, a system of two signs, a binary system, is in fact the fulfillment of our desire.

Two, it turns out, is also the least number of signs needed to transmit any information whatever. In a culture that abounds with vehicles of information it is not easy to imagine a situation where one is both starved for information and confined to the minimum number of signs. To conjure up such a setting, imagine Abe, a youngster in Missoula, Montana, in 1911, who has inherited from his father Dan, a Harvard graduate, an undying love for the recently founded Boston Red Sox. Abe has secured a schedule of their games, but has no radio or telephone available to find out how the Sox did on a particular day and depends on the belated reporting of the *Daily Missoulian* to keep up with his heroes. Imagine further that Abe's older brother John lives in Boston, and that John and Abe have conspired to prevail on two reluctant Western Union operators, one in Boston and one in Missoula, to provide, at the margin of legality and propriety, information about the Sox's game on the day in question.

The operators, reluctant as they are, agree to transmit just one signal on the day after the game, say the signal for the number one. Now what message could John convey to Abe? That the Sox had won? Unhappily that message would have been false half of the time. It was (almost) invariably true that they did play a game on the scheduled day. So that message could have been correctly sent with just one sign. But it would not have been news. It would have been uninformative. Abe, possessing a schedule, already knew when the Sox were to play. The conclusion is obvious. It takes at least two signs to convey any information. John and Abe, let us assume, succeed in securing a second sign, a zero, say, and John is now able to signal Abe whether the Sox have won or lost. A one means they won, a zero, they lost.

It is often thought that two signs also provide the elementary *measure* of information—the amount of information that is carried by one of two possible signals—by a binary (one of two) digit (signal). By happy linguistic accident, the basic *bit* of information can be thought of as a contraction of *binary digit*.[10] Departing from this presumably basic unit of information, Claude Shannon in 1948 gave an account of how to measure information and how to judge the fidelity and economy of communicating information. His paper, "The Mathematical Theory of Communication," became the landmark of information theory.[11] Its broader cultural significance and influence, however, have always consisted uneasily with its precision and formal elegance.

Shannon himself sought to confine his project to the technical problems of signal transmission. Not so Warren Weaver whose expository essay has accompanied Shannon's paper almost from the start. He acclaimed Shannon's theory with clashing cymbals:

> The obvious first remark, and indeed the remark that carries the major burden of the argument, is that the mathematical theory is exceedingly general in its scope, fundamental in the problems it treats, and of classic simplicity and power in the results it reaches.
>
> This is a theory so general that one does not need to say what kinds of symbols are being considered—whether written letters or words, or musical notes, or spoken words, or symphonic music, or pictures. The theory is deep enough so that the relationships it reveals indiscriminately apply to all these and to other forms of communication. This means, of course, that the theory is sufficiently imaginatively motivated so that it is dealing with the real inner core of the communication problem—with those basic relationships which hold in general, no matter what special form the actual case may take.[12]

Much excitement rippled across the scientific and cultural waters in the wake of Shannon's and Weaver's writings. Within a decade of their publication Yehoshua Bar-Hillel observed, "Papers on applications of Information Theory in psychology, linguistics, sociology, anthropology, physics, etc. appear in rapidly increasing numbers."[13] The excitement was fueled by a plausible analogy between the natural sciences and information theory. The latter, Donald MacKay claimed, "enables us to speak precisely and quantitatively. It provides objective substitutes for intuitive criteria and subjective prejudices."[14] Thus the theory of information, it appeared, at long last did for this crucial force of nature and culture what the sciences had done for matter, energy, and organisms. And just as scientific theories by way of technology dramatically advanced human health and comfort, so one could expect that information theory and technology would usher in a revolution of experience and creativity. Though the scientific excitement about information theory has run its course by now, the cultural and technological promise of the information revolution has only begun to crest.

The promise was that information theory would allow us to measure, control, and enhance information about reality and so enlarge and enrich the scope of human experience. The theory suggested that the value of information lies in its contingency, its unpredictability. To be told that the sun will rise tomorrow is to receive no information. To learn that one has won the jackpot in the lottery is to have great news. The trite, the hackneyed, the ordinary yield little information. What is rare, unlikely, surprising makes for much information. What information theory seemed to provide is a way of saying precisely just how little "little" and how much "much" information is. The more surprising a message, the greater the amount of information it contains; and the greater the variety of available signs, the more surprising a particular set of signs that conveys the message.

John, communicating with Abe, was confined to the minimal case of variety, to a choice between two signals. But now imagine that the Western Union operators in a fit of generosity made room not just for one of two digits but provided two places in succession so that John could mark the first place with one or zero depending on whether the Red Sox had won or lost and the second with one or zero depending on whether the immortal Tris Speaker had scored or not. A third place for a one or zero would be required to inform Abe about the scoring or not of the somewhat less immortal Harry Hooper.

Given one binary digit, John had a choice between two possibilities—a possibility space of two, we might say. Given two binary digits, the possibility space grew to four:

1. The Sox won, and Speaker scored.
2. The Sox won, and Speaker failed to score.
3. The Sox lost, and Speaker scored.
4. The Sox lost, and Speaker failed to score.

Given three places, John could select from eight possibilities.

Another way of looking at the guiding intuition of information theory is to think of the increase in available signs as corresponding to growing precision in the information conveyed. If the system of signs is a grid, then greater variety of signs corresponds to a finer grid and a finer grid to more precise information. If I tell you that the ancient camp of the Salish in the Rattlesnake Valley is located in section 25, the information you receive is relatively vague. If I divide each section into quarters and again locate the camp, you have more precise information. If I refine the grid twice more and select one of the resulting squares, you finally have the fairly precise information of the following legal description: Southwest One-Quarter Southwest One-Quarter Southwest One-Quarter, Section 25, Township 14 North, Range 19 West, Montana.

In the same way we can think of Abe as having been given, by his fairy godmother, a magic window that represents the world of the Boston Red Sox. Imagine a white square, a foot to the side, on the wall above Abe's desk. The minimal structure would be a division into two parts, the win side and the loss side. If the fairy's magic wand appears on the win side, Abe celebrates. After a week, the fairy divides each half once again into "Tris Speaker scored" and "Tris Speaker failed to score" halves. Now she has four possibilities and can place her wand more informatively. Things get better yet when she divides each of the four parts for Harry Hooper's scoring or not. Now there are eight possibilities.

Something like an analogy between more information and a progressively finer grid with increasingly precise selections was implied by R. V. L. Hartley. What are regions in a grid for us are symbol sequences for Hartley:

> By successive selections a sequence of symbols is brought to the listener's attention. At each selection there are eliminated all of the other symbols which might have been chosen. As the selections proceed more and more possible symbol sequences are eliminated, and we say that the information becomes more precise. For example, in the sentence, "Apples are red," the first word eliminates other kinds of fruit and all other objects in general. The second directs attention to some property or condition of apples, and the third eliminates other possible

colors. It does not, however, eliminate possibilities regarding the size of apples, and this further information may be conveyed by subsequent selections.[15]

One might conclude then that when the variety of signs available to John and Abe is two, the amount of information is two. When the variety of signs is four, a particular message is twice as surprising, and the amount of information is twice as large—four. If eight possibilities are provided for, a message is four times less likely and more surprising than in the initial case, and the amount of information is eight. Similarly, when the information about the location of the camp is determined by two possibilities (inside or outside section 25), the amount of information is two. When the section is divided into four times four times four squares and the camp is located in one of them, the information is sixty-four.

However, measuring information this way exposes an ambiguity in our intuitive notion of how to determine the quantity of a particular piece of information. To be confronted with ambiguities by a theory and to be forced to resolve them surely looks like a virtue in scientific theory; and, as MacKay promised, information theory "provides objective substitutes for intuitive criteria and subjective prejudices."[16] The ambiguity surfaces when you realize that if you can say a thousand different things on one page, you can say not just two thousand different things on two pages, but a thousand times a thousand; and a thousand times a thousand times a thousand, that is, a billion different things on three pages. On this view—the view we have taken of John and Abe's situation and of the location of the camp—the amount of information grows exponentially and explosively as the space for the vehicles of information is doubled, tripled, and so on. But this understanding collides with the equally or perhaps more plausible intuition that on two pages you can say twice as much and on three pages three times as much as on one. This latter position is in fact the more useful one and is now generally assumed in information theory and technology.

There is a lawful relation between the actual space that signs occupy and the possibility space of reference that is marked out by the signs. Thus when the space for binary digits that the Western

Union operators allot to John and Abe grows from one to two and then to three, the possibilities of reporting about the Boston Red Sox grow from two to four and then to eight. There is a function, a mathematical input-output device, that takes the possibilities of referring to reality as the input and gives you the number of binary digits needed to convey an actual reference to reality as the output. That function is the base-two logarithm.[17] Thus if the input is two (possibilities of reference), the base-two logarithm yields one (binary digit); for four (possibilities) it yields two (binary digits); and for eight (possibilities), three (binary digits). In this way we can express the possibilities of reference in the elementary measure of information, the bit.[18] When, for example, there are sixty-four possibilities of locating a camp, a report on a particular location conveys base-two-logarithm-of-sixty-four bits—six bits of information.

Given this apparent fit between the variety of signs and the contingency of things, it is plausible to believe that through the insight of information theory and the ingenuity of information technology, information systems would become ever more powerful and, as a consequence, reality in all its richness and contingency would become ever more commandingly present to human experience. To put this more precisely, one could expect that the increasing power and sophistication of the vehicles of information would bring in their train ever richer contents of information.

Content

Alas, it is confused at best and misleading at worst to assume that in information the linkage between signs and things can be rendered so clear and precise that the contingency of things can be measured by the variety of signs. And it is misguided to believe or merely to hope that larger and more sophisticated systems of signs will increasingly accommodate the presence of things. There are, to be sure, instances where the match between signs and things is so close that the number of bits is a measure of the contingency of things. But such cases, far from representing a natural harmony between information and reality, require us to confine and partition reality antecedently and artificially.[19] It is only when the world of the Boston Red Sox is first reduced and confined to a magic window of two, four, or eight segments and a corresponding number of possible events, that the number of bits matches the weight of what the bits are about. Similarly, the unfathomable richness of a Salish camp must be reduced to the issue of location within a predetermined grid before the number of bits corresponds to the precision of what we are told about the camp.[20]

The real world, however, does not submit to an antecedent confinement and partitioning into possible states or events. Bar-Hillel takes a hamlet as a little universe to clarify how we can hope to measure the content of information, that is, to measure not just the signs but what the signs are about. How rich is the reality of Bar-Hillel's little world? The hamlet has three inhabitants, and each has one of two properties, male or female, young or old.[21] And yet the choice of so austere a picture was prudent because it kept the possible states of the hamlet surveyable. Had Bar-Hillel chosen a hamlet of a hundred people and enlarged the sets of possible traits of each person from four to eight, the number of possible states of that hamlet would have been astronomical, and we still would have no means of referring to the streets, the houses, or the businesses of the hamlet.

There is no lawful relation between the number of bits and what they are about. A person's happiness can hang by one bit as it would in the olden days when it was the answer to the question "Will you marry me?" and as it still does today when one bit, usually smothered in redundancy, tells you whether or not you got the job of your dreams. In such cases, the context of a person's life comes to be focused into a fateful disjunction and question: Will it be this or that? And the answer constitutes one enormously significant bit of information.

Reality does not often assume such a sharply etched outline of two possibilities, and one bit of information rarely comes to be the vehicle of resolution. Most often a bit of information is about next to nothing and so are all too often hundreds, thousands, millions, and even billions of bits. It is evident from such observations that we are never totally in the dark about the content of information that is conveyed by a particular set of signs. We can talk approximately and generically about the value of information content. Critics of literature and poetry make it their business to do so. But there is no rigorous way of linking content to signs short of imposing a rigid structure of predictability on the contingency of things. The content or weight or significance of information depends not only on the nature of the things that signs are about, but also on the context of things and the intelligence of the recipient of the information.

The *bit of information* in its most austere sense is a measure of information space, and by itself the number of bits of a set of signs tells us nothing about whether or how the space has been filled with content. The *bit* is a technically useful and even crucial notion. But it is hard to stick to its clear and defensible meaning.[22] Shannon emphatically restricted himself to the technical problems of signal transmission and yet invited confusion when he proposed (following Hartley) that the variety of signs or signals (of "messages" as he puts it) "can be regarded as a measure of the information produced when one message is chosen."[23] Today's usage of the *bit* ranges from the sober and technical to the confused and extravagant. Engineers refer to the billions of bits on the hard drive of a computer as disk space rather than information. Enthusiasts of the information revolution, to the contrary, easily slide from the fact that we can store and manipulate millions and billions of bits to questionable assertions on how radically and beneficially information storage and processing will transform our experience of reality.

Though information theory and technology do not by themselves support the claim that they will greatly enhance *information about* reality, they may more plausibly encourage the hope that a precise understanding of the possibility space of information would put at our disposal powerful *information for* the shaping of reality and enhance our freedom of personal expression and artistic creativity. A hint of that hope may be seen in Weaver's pronouncement that "this word information in communication theory relates not so much to what you do say, as to what you could say. That is, information is a measure of one's freedom of choice when one selects a message."[24]

The hope that a truly unencumbered and clearly structured possibility space would engender a burst of actual creativity is not a prospect without historical support. In music, the steps of the natural scale of tones characteristically differ from one another in height—a pattern that is easily accommodated by voices and strings. If a keyed instrument such as an organ is tuned to these natural intervals, it sounds harmonious in certain keys but is also confined to that particular region of tones and begins to sound out of tune when it ventures into the distant keys.

Around Bach's time various efforts were made to temper the steps differently (to temper is to tamper) and to find tunings that would enlarge the harmonic space of harpsichords and organs. Bach was among those who took the ultimate step of making all notes nearly equidistant.[25] Here, as in the design of analytic geometry, it took a certain recklessness and rigor to impose on the ancient acoustic landscapes a uniform and unvarying measure. The reward was an evenly structured and totally accessible space of modulations and transpositions. But what impresses the cultural imagination most was that Bach not only helped to open a new possibility space, but appropriated and inscribed it in masterful fashion, most systematically in the forty-eight pieces of *The Well-Tempered Clavier* where he twice traversed all the major and minor keys of the twelve-tone scale, and most widely in his organ music where he frequently ranged over regions that are open only on a (nearly) equally tempered organ.

Information technology, however, has enlarged the space of our choices to an extent where it has lost all structure and resistance. Today's paradigmatic field of possibilities is not circumscribed by six or so octaves and twelve keys, nor is it a magic window of two or four or eight segments. It is more like the standard computer screen of the early nineties that had 640 times 480 pixels. Each pixel could have any one of 256 colors. Thus the screen has $256^{640 \times 480}$ or 1.3×10^{7398} different possible states. Given that there are "only" about 10^{78} protons and neutrons in the entire visible universe, the number of possible screen states is unimaginably large. Of course, no one has literal control of the entire range of possibilities, nor is anyone able to tell each from all the others. Still a consideration such as this along with a reminder of the tremendous growth in bandwidth for transmitting, in capacity for storing, and in power for processing information suggests that freedom of choice today is as likely stifling as liberating.

The force of reality does not naturally present itself in bits. But if we can theoretically grasp the structure of information, it is technologically possible to capture the surface and anatomy of reality by assigning bits of information to the facets and ligaments of things, and in this way information about, for, and as reality can be structured in bits with powerful results. "Bits of information" has therefore a broad generic meaning that is as useful in discussions of information technology as it is treacherous in discussions of contemporary culture. In any case, to understand how the characteristic bits of present-day information have insinuated themselves into our lives, we need to grasp the basic patterns in which bits of information can be combined with one another.

CHAPTER ELEVEN

Basic Structures

Division

The least and most basic structure of all seems to be division, a distinction when made by the mind, a difference when found in reality. Division, moreover, appears to be a principle of generation as well as the basis of structure. In Genesis, God's first creative act amounts to a gesture of dividing:

> In the beginning God created the heaven and the earth.
> And the earth was without form, and void; and darkness was upon the face of the deep. And the Spirit of God moved upon the face of the waters.
> And God said, Let there be light: and there was light.
> And God saw the light, that it was good: and God divided the light from the darkness.[1]

Here on earth, there is at least a spark of divine creativity when someone makes a difference. Leibniz thought as much when he rediscovered the binary system. It seemed to him the very image of creation. *Imago Creationis* is what he wanted to have inscribed on the commemorative medallion. Binary notation, in Leibniz's view, demonstrates that out of divine unity (one) and formless nothing (zero) everything could be generated. And there is more: "It is no less remarkable that there appears therefrom, not only that God made everything from nothing, but also that everything that He made was good; as we can see here, with our own eyes, in this image of creation. Because instead of there appearing no particular order or pattern, as in the common representation of numbers, there appears here in contrast a wonderful order and harmony which cannot be improved upon."[2] Leibniz went on to show how the sequence and structure of binary numbers can be generated by following a simple algorithm, and he was so impressed with this coincidence of order, beauty, and divinity that he sought to bring it to the attention of the emperor of China, "a lover of the art of arithmetic," because it "might serve to show him more and more the excellence of the Christian faith."[3]

A more sober and secular version of division was developed early in the twentieth century by the linguist Ferdinand de Saussure, who took the notion of difference to be necessary and sufficient for the structure of language. "Everything that has been said up to this point boils down to this: in language there are only differences. Even more important: a difference generally implies positive terms between which the difference is set up; but in language there are only differences *without positive terms*."[4] At a more speculative and sublime level, the philosopher Martin Heidegger in the late fifties hoped to illuminate the relation between humanity, reality, and technology through reflections on identity and difference.[5] Most boldly, perhaps, the physicist John Archibald Wheeler has conjectured that reality is the response to distinctions we humans propose, a hypothesis he has dubbed

> *It from bit.* Otherwise put, every it—every particle, every field of force, even the space-time continuum itself—derives its function, its meaning, its very existence entirely—even if in some contexts indirectly—from the apparatus-elicited answers to yes-or-no questions, binary choices, *bits.*
>
> It from bit symbolizes the idea that every item of the physical world has at bottom—at a very deep bottom, in most instances—an immaterial source and explanation; that which we call reality arises in the last analysis from the posing of yes-no questions and the registering of equipment-evoked responses; in short, that all things physical are information-theoretic in origin and this is a *participatory universe.*[6]

The first to realize the connection between difference and information technology was Charles Babbage, who in 1822 built an experimental calculator he called a "difference engine." He went on to design and started to build Difference Engine No. 1 and, when those efforts collapsed, made twenty drawings of the more advanced and elegant Difference Engine No. 2. Babbage's difference is not binary. It pertains to the method of finite differences that reduces polynomial functions to a series of subtractions and additions. Though Babbage used a system of ten rather than two digits, the more general notion of digitality with its precision and reliability was magnificently embodied in the bronze, iron, and steel of the Difference Engine. A model of No. 2 was successfully built in 1990–91, and one of its builders, Doron D. Swade, concluded his account of the venture with these ringing words: "Difference Engine No. 2 stands as a splendid piece of engineering sculpture, a monument to the rigorous logic of its inventor."[7] There is no denying the beauty of the machine nor its utility had it been completed in Babbage's lifetime. Today, however, it is a slow and lumbering computer. To make sure that the machine could be operated by hand, its builders added a four-to-one reduction gear to the handle, making it four times easier to turn and four times slower to operate. To realize complex mathematical operations in a machine, something swifter than Newtonian wheels, cams, racks, and levers was needed. That something was the electron, whose current moves nearly at the speed of light.

Vacuum tubes were the first electronic devices to harness electrons for switching.[8] In 1944 and 1945, John Eckert, Herman Goldstine, and John Mauchly assembled 18,000 of them, together with 70,000 resistors, 10,000 capacitors, and 6,000 mechanical switches into the ENIAC, the Electronic Numerical Integrator and Computer. It could do five thousand additions or subtractions in one second and perform calculations in one minute that would have required twenty-four hours of a human equipped with a mechanical calculator.

Like light bulbs, vacuum tubes fail. In the ENIAC, one tube per day failed on the average, and it took an hour to find and replace it.[9] Obviously the failure rate would rise with the size of the computer, and the time for uninterrupted computation would shrink. The ENIAC was a behemoth already. It was a hundred feet long, ten feet high, three feet deep, and weighed fifty tons. For computers to grow in power, a simpler and more reliable electronic switching device than the vacuum tube was needed—the transistor, invented in 1948 by John Bardeen, Walter Brattain, and William Shockley. In 1959, Jean Hoerni etched a transistor on a silicon wafer.[10] Five years later, Gordon Moore realized that the number of elements on a silicon chip had been doubling every year and a half, a trend that has continued to this day and has come to be known as Moore's Law. Today, a typical chip like the Pentium contains over three million transistors.

The silicon chip was the medium that joined the elementary unit of information with the elementary particle of the electron and lent the primal divisions—light and dark, one and zero, yes and no, true and false, high and low, open and closed—the speed, complexity, and reliability that have transformed our sense of information and reality. "Computer" is what we call this multiple juncture generically. Though its structural basis is of crystalline clarity, its cultural status has become ever more hazy. What is a computer taken as a cultural force? Richard Coyne has urged that the prominent metaphors that have served to insinuate the computer into our culture are the computer as calculating device, as drawing tool, and as intelligence. The first two metaphors—"the computer as an adding machine, slide rule, ledger, codebook" and "the computer as drawing surface and pen, a dynamic chart"—remind us that, like cultural information, technological information raises the function of its predecessors to an entirely new level.[11] Computers furnish vastly increased information *about* and *for* reality. It is as intelligence that computers and the technological information they embody step forward *as* reality in their own right.

There is, however, one feature that the three computer metaphors have in common. Whatever else computers may be in our culture, they are automatic in the sense that they do things for us that we cannot or do not want to do ourselves. To be sure, rather few computers are entirely automatic, excluding human agency altogether. But all of them cushion and comfort the human condition. In some way they disburden us from having to cope with the contingency of reality.

Boolean Algebra

To cope is to make a beneficial difference. A trap is a device that makes such a difference as regards the contingencies of rodent behavior. In fact it makes the simplest possible difference. It responds to one of two states, pressure or no pressure on the trigger, and reacts in one of two ways, it is open or not. We can diagram the behavior of a trap this way:

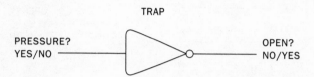

TRAP

PRESSURE?
YES/NO

OPEN?
NO/YES

When the answer to PRESSURE? is YES, the answer to OPEN? is NO; when the answer to PRESSURE? is NO, the answer to OPEN? is YES.

Complexity of course is the soul of computing. As a first step toward understanding complex automatic devices, take the weather as the great contingency and a window opener as a device for coping with it. The opener responds to two alternatives, to whether it is dry or not and to whether it is calm or not. Assume you have a daughter who is a lover of fresh air and wants her window open unless it is windy and raining. We can diagram this window opener thus:

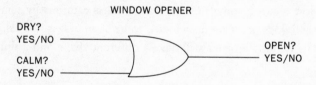

WINDOW OPENER

DRY?
YES/NO

CALM?
YES/NO

OPEN?
YES/NO

It is obvious, however, that this one diagram fails to show just how the device works. To be clear, we need to show the four possible states the opener can be in.

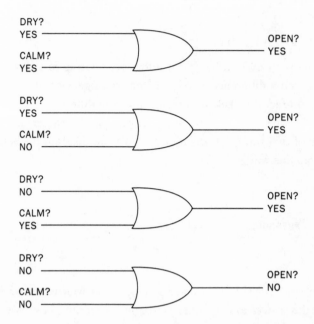

These four possibilities can be listed more compactly in a table:

DRY?		CALM?		OPEN?
YES	combined with	YES	results in	YES
YES	combined with	NO	results in	YES
NO	combined with	YES	results in	YES
NO	combined with	NO	results in	NO

Now assume you also have a son who is cautious by temperament and does not want his window to be open unless it is calm and dry. His window opener will have a different shape and a different table.

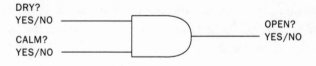

DRY?		CALM?		OPEN?
YES	combined with	YES	results in	YES
YES	combined with	NO	results in	NO
NO	combined with	YES	results in	NO
NO	combined with	NO	results in	NO

These window openers have important features in common with their hypertrophic cousins, the computers. At the beginning and the end, where they interact with the world, they have sensors to respond to the world and effectors to react to the world. The window opener will have a hygrometer to measure moisture and an anemometer to measure air movement, and it will have a motor to open and close the window. Every computer has such input and output devices. In a notebook computer used for word processing, the keys and the mouse are typically the input devices, and the screen is the output device. The writer embodies the unpredictable contingency of the world. Between input and output, however, there is nothing but the pure structures of yeses or noes. Even a trap implicitly represents such a structure, as is evident in this table:

PRESSURE?		OPEN?
YES	results in	NO
NO	results in	YES

The stability of these internal structures rests entirely on mere and pure difference. The two kinds of inputs, the yeses and noes, must be kept clearly distinct. But what you call them does not matter. Instead of YES and NO, you could use TRUE and FALSE, 1 and 0, HI and LO, or whatever. TRUE and FALSE could easily stand in for YES and NO when it comes to the window opener for the cautious:

DRY?
TRUE/FALSE

CALM?
TRUE/FALSE

OPEN?
TRUE/FALSE

If it is TRUE that it is DRY AND it is TRUE that it is CALM,
then it will be TRUE that the window is OPEN.

If it is TRUE that it is DRY AND it is FALSE that it is CALM,
then it will be FALSE that the window is OPEN.

If it is FALSE that it is DRY AND it is TRUE that it is CALM,
then it will be FALSE that the window is OPEN.

If it is FALSE that it is DRY AND it is FALSE that it is CALM,
then it will be FALSE that the window is OPEN.

In the inner structure of the cautious person's window opener, the "combined with" of the yeses and noes, corresponds to an AND between TRUES and FALSES. In the structure of the fresh air lover's window opener, the connector that would properly combine the TRUES and FALSES would be an OR, more particularly the inclusive OR we sometimes render as AND/OR. Accordingly, computer scientists call this

an AND gate, and this

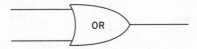

an OR gate, and this

a NOT gate, or an inverter. Each gate has its characteristic table:

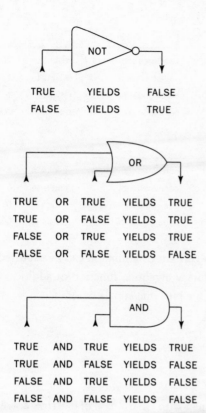

		NOT	YIELDS	FALSE
TRUE			YIELDS	FALSE
FALSE			YIELDS	TRUE

TRUE	OR	TRUE	YIELDS	TRUE
TRUE	OR	FALSE	YIELDS	TRUE
FALSE	OR	TRUE	YIELDS	TRUE
FALSE	OR	FALSE	YIELDS	FALSE

TRUE	AND	TRUE	YIELDS	TRUE
TRUE	AND	FALSE	YIELDS	FALSE
FALSE	AND	TRUE	YIELDS	FALSE
FALSE	AND	FALSE	YIELDS	FALSE

These structures mesh well with our intuitions. To satisfy an AND, much has to be true, to satisfy an OR, little.[12] The fresh air lover is tolerant; her window is open most of the time; she is an OR person. The cautious person is demanding; his window is rarely open; he is an AND person. She will often say ". . . OR whatever," while he says, ". . . AND one more thing."

However, their elective affinities with features of the real world should not distract us from the astounding generality of these binary structures. If you replace the yeses and noes of the fresh air lover's and the cautious person's window openers with ones and zeros, you get this table:

1	combined with	1	results in	1
1	combined with	0	results in	1
0	combined with	1	results in	1
0	combined with	0	results in	0

CAUTION

1	combined with	1	results in	1
1	combined with	0	results in	0
0	combined with	1	results in	0
0	combined with	0	results in	0

Evidently, the FRESH AIR combination can be captured by something close to the ordinary algebraic function of addition (except for the stipulation that one plus one equals one):

$$1 + 1 = 1$$
$$1 + 0 = 1$$
$$0 + 1 = 1$$
$$0 + 0 = 0$$

And the "combined with" in the case of CAUTION corresponds to multiplication:

$$1 \cdot 1 = 1$$
$$1 \cdot 0 = 0$$
$$0 \cdot 1 = 0$$
$$0 \cdot 0 = 0$$

The structure of these binary functions was discovered by Babbage's contemporary, George Boole (1815–64), and is commonly known as Boolean algebra. Boole's intention had been to reduce reasoning to algebra. Obviously the connective OR is related to TRUE and FALSE the way Boolean "plus" is related to one and zero:

TRUE OR TRUE = TRUE	$1 + 1 = 1$
TRUE OR FALSE = TRUE	$1 + 0 = 1$
FALSE OR TRUE = TRUE	$0 + 1 = 1$
FALSE OR FALSE = FALSE	$0 + 0 = 0$

Similarly AND and "times" go together (unhappily and confusingly AND does not go with "plus" or "and" in Boolean addition; it goes with "times" and multiplication):

TRUE AND TRUE = TRUE $1 \cdot 1 = 1$
TRUE AND FALSE = FALSE $1 \cdot 0 = 0$
FALSE AND TRUE = FALSE $0 \cdot 1 = 0$
FALSE AND FALSE = FALSE $0 \cdot 0 = 0$

How does all this apply to reasoning? To reason, in the logician's sense, is to extract information from information according to structural rules. You look from the carriage house at your daughter's window. It is open. You know that it is only open if it is dry and/or calm. You can see that it is windy, but you are not sure whether it is raining. Now you reason as follows:

1. It must be DRY AND/OR CALM (because the window is open).
2. But it is NOT CALM.
3. Hence it is definitely DRY.

This piece of reasoning logicians call a disjunctive syllogism. What makes it logical reasoning is that if (1) and (2) are TRUE, the truth of (3) follows mechanically, by the laws of Boolean algebra alone. In effect you are given an equation with one unknown, and Boolean algebra tells you that the value of the unknown must be 1 or TRUE:

DRY?		CALM?		WINDOW OPEN?
?	+	0	=	1
?	OR	FALSE	=	TRUE

DRY is 1 or TRUE.

A fallacy is the failure to obey structural rules. Let us assume you reason thus:

4. The window is open (equivalent to: it must be dry AND/OR calm).
5. It is calm.
6. Therefore it must be dry.

Expressed in Boolean algebra that would be:

DRY?		CALM?		WINDOW OPEN?
?	AND/OR	TRUE	=	TRUE
?	+	1	=	1

Is DRY TRUE or FALSE? Is it 1 or 0? There is no way of knowing from (4) and (5) alone, and this is reflected by Boolean algebra in that either 0 or 1 (remember the stipulation $1 + 1 = 1$) would make ? + $1 = 1$ TRUE. Hence by Boolean algebra alone, you cannot say whether DRY is 1 or 0, TRUE or FALSE; to claim it is true would be rash and fallacious.

Are Boole's two binary operations the minimum that is needed to generate all binary structures?[13] Logicians and engineers will give different answers. The logician Henry Sheffer showed in 1913 that one operation is sufficient.[14] It corresponds to the English connective "neither . . . nor . . ." and yields a true proposition only when you take two false propositions and connect them with "neither . . . nor . . ." It is false that it is snowing red roses; and it is false that it is raining cool wine. Hence it is true that it is neither snowing red roses nor raining cool wine. All other combinations of true and false components yield false compounds. Engineers call this a NOR gate, and it has this table:

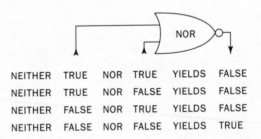

NEITHER	TRUE	NOR	TRUE	YIELDS	FALSE
NEITHER	TRUE	NOR	FALSE	YIELDS	FALSE
NEITHER	FALSE	NOR	TRUE	YIELDS	FALSE
NEITHER	FALSE	NOR	FALSE	YIELDS	TRUE

The NOR gate, alone or in combinations, does everything the other gates do.

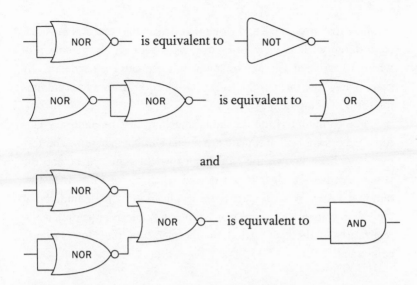

and

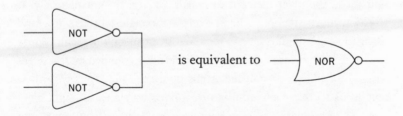

To an engineer this is an awkward way of making a simple point. More incisively perhaps, the engineer would argue that a NOR gate, though sufficient as a basis, is not basically simple since it can be put together from the structurally simpler NOT gates because

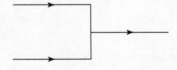

The logicians, however, will reply that a juncture like this

is equivalent to a Boolean product or a logical conjunction and that the engineer, logically speaking, is using two gates, a NOT gate and an AND gate.[15] To the philosopher, as distinguished from the logician, the engineer's position is most pleasing. It gives pride of place to the simplest structural device, the one that makes the least and most basic difference. This view, of course, must rely on the geometry of wiring to arrive at AND and OR gates. But geometry is needed in any case. Being "the form of the world," as Gordon Brittan has said, it is the form of computers as well.[16]

The computer is the millennial machine of contemporary culture. Syllogisms and automata are ancient in comparison. The former have been known for more than two thousand years, and the latter have been built in sophisticated versions for more than two hundred years. What is new is the explication of their underlying logical structures as binary operations of Boolean algebra and the realization of these operations in the electrons that flow through computers.

Transistor and Computer

The fundamental device that provides for the transition from structure to matter is the transistor, and for the last four decades the favored stuff to materialize the structures in has been silicon, the second most abundant element on earth. Silicon occurs naturally as silica or, more precisely, silicon oxide—the material rocks and sand are made of. When oxygen is removed from silica, the result is pure silicon. The nucleus of an atom of silicon is surrounded by four electrons. In a silicon crystal, the atoms pool their electrons with their neighbors so that each atom is surrounded by a shell of eight electrons. The regularity of common electrons knits silicon atoms into the three-dimensional lattice we call a crystal.

Yet the same regularity also anchors the electrons in their appointed places and makes the movement of electrons that is crucial if they are to carry information difficult to effect. To provide for mobility, the crystalline structure of silicon is carefully rearranged by putting here and there one phosphorus atom in place of a silicon atom. Phosphorus atoms have five electrons rather than the silicon four, so one orphaned and footloose electron remains in the crystalline arrangement. Similarly, a silicon crystal can be doped with boron atoms which have three electrons and leave an empty place or hole in the lattice of nuclei and electrons. Voilà, a semiconductor—a material that conducts electricity when an electric charge is applied to it and makes electrons move.

Now imagine two pieces of phosphorus-doped silicon inserted in a substrate of boron-doped silicon and connected to aluminum wires. One of the pieces—we will call it the "source"—is at a potential of zero volts. The other piece—the "drain"—is given a positive potential and hence is attractive to the electrons in the source. But the electrons cannot cross the boron-doped region that separates source and drain. Next to this area, there is a layer of silicon dioxide, a good insulator, and attached to it is another aluminum wire—the "gate" (not to be confused with a logic gate). Because of the insulating silicon dioxide, the gate is not directly connected with the semiconductors to its right. But it can influence them through the electrical field it emits when it is electrically charged.

So far everything is quiet. No electrons are flowing. The transistor is OPEN, or OFF. But now a positive potential is applied to the gate. Its electric field attracts the electrons in the substrate to a thin layer right next to the gate. To the right of this layer of electrons there is an electronic vacuum that serves as a channel, and now electrons are flowing from the source to the drain. The transistor is CLOSED, or ON.

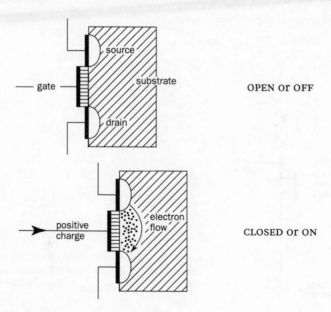

OPEN or OFF

CLOSED or ON

Reduced to its schematic essentials, a transistor is rendered thus:

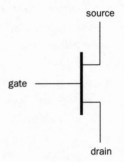

source

gate

drain

It represents a switch.

But how do we get from a switch to the simplest logical device, an inverter or NOT gate? We connect a resistor to the source, a resistor being rendered as a zig-zagging line. To a spot between the resistor and the source we connect a wire we call "output." The gate wire we call "input." When the input voltage is LO, the supply voltage to the (inactive) resistor and source remains HI and is therefore HI at the output. When the voltage or potential to the input or gate is HI, the transistor is closed, a current flows, the current activates the resistor, the resistor resists, and the voltage drops to LO and is registered as LO at the output (see fig. 14).

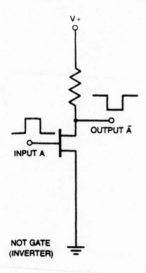

V+

OUTPUT Ā

INPUT A

NOT GATE
(INVERTER)

FIGURE 14 A NOT gate (inverter), from William C. Holton, "Large-Scale Integration of Microelectronic Circuits," in *Microelectronics* (San Francisco: Freeman, 1977), 29

Thus we get a familiar table:

INPUT?	OUTPUT?
HI	LO
LO	HI

For purposes of function and design, only the input and output wires and a distinctive symbol between them matter. Hence the shape of the inverter:

The triangle is a box that contains a resistor and transistor. We can now design gates at the level of inverters whose innards no longer concern us. We can build a NOR gate this way:

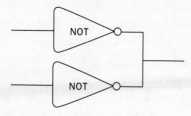

And an OR gate thus:

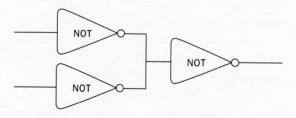

We can, if we wish, open the inverting boxes and look at the resistors and transistors that are at work here (see fig. 15). But we can also put the three inverters of the OR gate in one box and get this:

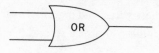

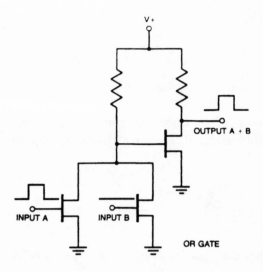

FIGURE 15   An OR gate,
from William C. Holton,
"Large-Scale Integration of
Microelectronic Circuits," in
Microelectronics
(San Francisco: Freeman,
1977), 29

We can go on from there and combine NOT, OR, and AND gates into
something that functions as an adder (fig. 16). If we gather many
such arrays into still larger boxes and label and connect them appropriately, we get the outline of a simple computer's heart and soul, the
central processing unit, or CPU (fig. 17). This is in fact the outline
of the CPU on Intel chip 8080, the one that prompted Bill Gates and
Paul Allen to write software for a living and to found Microsoft.[17]

What then is a computer from the standpoint of basic structures?
It is Boolean algebra in motion. If we could make the 8080 gigantic
and transparent, if we could slow it to a snail's pace and color the
electrical impulses red, then we would see a crimson stream emerge
from the clock (timing and control) that would travel along some of
the lines and through some of the boxes or, on a smaller scale,
through the wires and logic gates, and come to rest in buffers,
latches, and registers. Another stream of pulses would follow and
flow and rest in a different pattern. The pulses of red would correspond to ones, the absence of pulses to zeroes. The rhythmically
changing patterns of pulses represent the process of information processing. Or if you like an acoustic simile, think of the patterns of
pulses as chords that the machine takes from the score of the program, sends through its wires, and modulates through its logic gates,
in a rhythm of two million chords a second.[18]

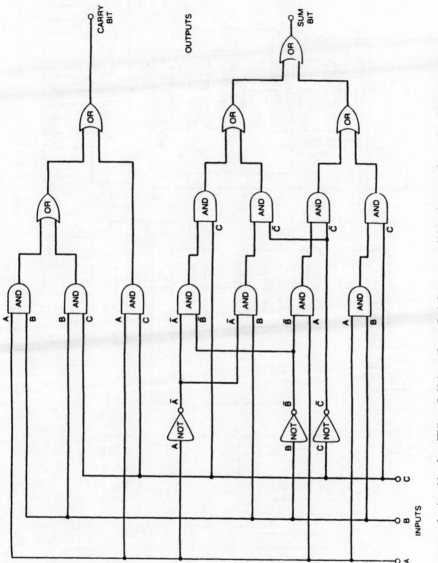

FIGURE 16 An adder, from William C. Holton, "Large-Scale Integration of Microelectronic Circuits," in *Microelectronics* (San Francisco: Freeman, 1977), 32

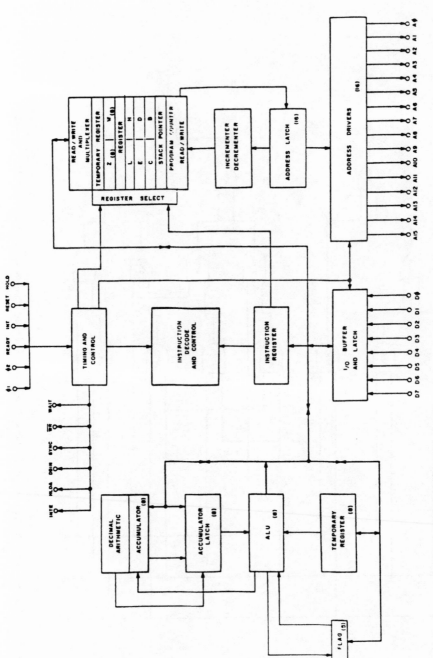

FIGURE 17 The central processing unit (CPU) of the 8080 processor chip, from Edward Roberts and William Yates, "Altair 8800," *Popular Electronics*, January 1975

The more closely you look at a computer chip, however, the more remote its function becomes. At the level of electron streams you could no longer tell an AND gate from an OR gate, far less could you determine whether you were looking at a chip that was programmed to solve equations or tell lies. One has to rise to a level of observation distant enough to reveal the larger functional parts of the chip. Like Babbage, we can think of a computer as a processing plant. What it processes is information. It stores its material—the data—in a store today called RAM (random access memory), and processes it by means of a mill—the CPU. There will also be a funnel to get the material into the store—the input device—and a chute where the processed material emerges—the output device. The crank of the mill that puts the machine through cycle after cycle corresponds to the computer's clock.[19]

What makes the computer such a powerful information mill is the fact that its store contains two kinds of materials, stuff that is to be processed (data) and stuff that rearranges the gears of the mill (instructions). In the computer memory, instructions and data are carefully layered as on a list, and as the computer goes through its cycles and down the list, it alternates between processing data and following instructions to rearrange itself for the next batch of data. The set of instructions in memory is the program, and a computer strictly so called not only rearranges itself according to its program, it allows you to rearrange the program itself.

There is a special delight in playing with the tremendous power of a computer, in taking a problem apart, reducing it to a list of instructions and data, and, with the pressing of a key, making the program become a physical event that busily and unerringly goes through its paces and offers up its results. Consider Pythagorean triples. For a right-angled triangle, the sum of the squares on the shorter sides equals the square of the longest side. This is generally captured in the familiar equation:

$$x^2 + y^2 = z^2$$

The equation, as we know, comes out right for 3, 4, and 5:

$$3^2 + 4^2 = 5^2$$

Are there other groups of three numbers that fit the equation? You can write a program that combs the natural numbers from the bottom up. Press the key that gets the program running, go get a cup of coffee, and when you come back there will be a list, depending on the power of your computer, of a few, a few dozen, or a few hundred Pythagorean triples.

In such an experience the creative alchemy of difference and the sober clarity of structure seem to coincide. Just as the philosopher's stone would turn base metals into gold, so the silicon chip transmutes vague and fallible reasoning into a perspicuous and reliable operation. Mind becomes matter, information becomes structure. Here the computer moves into the penumbra of Richard Coyne's third metaphor—the computer as intelligence. Yet however magical it may seem, we can soberly satisfy ourselves that the whole is comprised of nothing but physical particles, lawfully ordered. Unlike Johann Nepomuk Maelzel's chess automaton that had to contain and conceal a human being in its recesses if it was to work, a silicon chip harbors neither a soul nor a secret.[20]

The sense of power and delight that is engendered by the programmable computer began to inform the everyday world of this country when in 1975 the first affordable home computer with the then most advanced computer chip came on the market.[21] The Altair 8800 cost $397 as a kit and $498 assembled and tested. Its design instilled in its user an acute awareness of a computer's basic structure and of how one builds a program from the ground up. The Altair had a row of sixteen two-state switches (represented in the lower right corner of the CPU diagram in fig. 17) that had you spell out in binary digits the instructions to the computer, the data to be operated on, and the locations of instructions and data in the computer's memory. It had two rows of light-emitting diodes (LEDs) that displayed results as binary numbers. Thus you were instructed to

> store the load accumulator instruction at location 0 by using the binary number for 58 (00111010). Set this binary input up by using switches D0 through D7, with a 1 represented by the switch in the up position and a 0 with the switch in the down position. Hence the switch sequence for 00111010 would be: D7 down, D6 down, D5 up, D4 up, D3

up, D2 down, D1 up, D0 down. Store this number at location 0 by operating the DEPOSIT switch. The D0 through D7 LEDs should now match these settings with a lighted LED indicating a 1 and a darkened LED indicating a 0. None of the A0–A15 LEDs should be on indicating location 0.[22]

And so on until you were ready to operate the RUN switch and await the result as a row of light and dark LEDs.

The early computer aficionados and aficionadas continued to derive satisfaction from looking all the way down to the fundamental structure of electronics and Boolean algebra even when communication with the computer proceeded by way of keyboard, screen, and a kind of language that bordered on ordinary English, where to put a number N into memory you would type "INPUT N" and to run a program, "RUN."

For technically inclined people, the early computers presented a microcosm of technology that they were able to comprehend and inhabit in its entirety, unlike the world of their work where they were confined to a small niche and unlike society at large that alienated them through its forbidding and irrational complexity.[23] For technically untutored or timid persons, the early computers presented an opportunity to become acquainted and easy with some of the complexities of mathematics, science, and engineering and so to acquire a franchise in the dominant and distinctive culture of our time.

The clarity of vision and sense of power centered in the computer appeared to radiate powerfully into the world of the early computer users. The computer promised to bring transparency and control to the dark and messy corners of the domestic sphere. People used their computers to catalog their books, LPs, recipes, and tools. They used them to straighten out their budgets, investments, and taxes. Reaching outward from their homes, they reduced their address books and telephone numbers to computerized lists. And reaching further yet, they joined user groups and by way of telephone lines formed networks.

Most ambitiously, the early computers inspired a vision of a new commonwealth. Computers would break through the walls of privileged and distant information and usher in an age of equality and openness.[24] Communication and shared information were to nurture community, individual command of information would enhance autonomy. Computers would be the vehicles to rally the new sense of community and autonomy on behalf of ecology by confronting people with the distress of the environment and offering them salubrious alternatives to waste and destruction. Obviating the need to truck information and chauffeur people hither and yon, they would allow for a beneficial reordering of patterns of work and of urban space.[25]

The sense of wholeness and intimacy that early computer users felt was like the sense of renewal city dwellers may feel when in the summer they spend a month in a log cabin and learn firsthand what it is to haul water, build a fire, prepare a meal, and live in a surveyable and comprehensible world. Technological information seemed to recapitulate ancestral information. And what happened to this vision of coherence and this surge of competence? They were undone by the advancement of the millennial machine itself. At the time, chips like the Intel 8080 were the pinnacle of progress. In 1977, Robert Noyce compared them to the ENIAC and said, "An individual integrated circuit on a chip perhaps a quarter of an inch square now can embrace more electronic elements than the most complex piece of electronic equipment that could be built in 1950. Today's microcomputer, at a cost of perhaps $300, has more computing capacity than the first large electronic computer, ENIAC. It is 20 times faster, has a larger memory, is thousands of times more reliable, consumes the power of a light bulb rather than that of a locomotive, occupies 1/30,000 the volume and costs 1/10,000 as much."[26] The Intel 8080 was, within twenty years, followed by the 8088, 80286, 80386, 80486, and the various Pentiums. If we take an advanced version of the Pentium as today's standard, the number of transistors has grown by a factor of fourteen hundred, the speed of a CPU has increased by a factor of two hundred, and the Pentium is often matched with a system memory that exceeds the one in the Altair by a factor of 256,000. What does that amount to?

Perhaps the quality of this advancement can be likened to the step from a log cabin to a skyscraper, a move from a limited and laborious environment to a world of tremendous capacity, convenience, and speed. Many of the intimate engagements that a log cabin enforces are not just unnecessary but entirely impossible in a high-rise building. You simply cannot take the stairs to your office, haul all the water you need, make your own heat, nor even open a window. Like the coherence and intimacy of the log cabin, the transparency and comprehensibility of the Altair 8800 are distant memories.

Chapter Thirteen

Transparency and Control

Perspicuity and Surveyability

Historically considered, information technology (IT) has resulted from the convergence of two technologies: the transmission of information and the automation of computation.[1] Today IT is the common term for the ensemble of hardware and software—of computers, peripherals, and communication links on the hardware side, and of operating systems, programs, and data on the software side. Technological information could simply be defined as the object of information technology. But we can be more explicit and define it structurally as the information that is measured in bits, ordered by Boolean algebra, and conveyed by electrons. It has a plausible claim to representing the fundamental and universal alphabet and grammar of information. Surely it is remarkable that combinations of two signs and one or two operations are powerful enough to do the most complex calculations, contain and control music, capture and manipulate images, process words, steer robots, and guide missiles. Semantically, technological information holds the promise that, if properly linked with reality on the input side, the rigor of its algebra will faithfully preserve and process meaning and yield reliable and valuable information on the output side.

When I use the word "technological" to characterize information, I refer of course to a special and restricted meaning of the term. In a wider sense, cairns, tallies, clay tokens, and letters represent technologies as well. Here I use the word "technological" to refer to modern and in fact to the most recent technology of information. Not that its present and particular features, however philosophically pleasing, are necessary to something like technological information.

If electricity were biased toward three distinct states, we would measure information in tits rather than bits, and we would use a three-valued algebra to structure it.[2] Technological information, moreover, is conveyed by photons as well as electrons, and some day information may be processed by means of proteins, photons, or quantum effects.[3] What will remain are digital rigor, the massive logic and data structures, and the rapid processing of technological information.

None of the "pretechnological" kinds of information combined all three features. Writing and printing are digital, to be sure, albeit on a vicenisenary (twenty-sixfold) rather than binary base; and there are certainly gigantic information structures laid down in print or writing. But processing traditionally had to be done in the slow and laborious human way. Analog information contained in photographs and on maps, film, tapes, and vinyl records is more massive yet than written or printed information, and it can be handled a little more easily and quickly by means of the traditional editing and displaying devices. But like a satellite in low orbit, analog information is in constant danger of falling back into reality. Every time it is copied or displayed, it suffers irreversible damage. Its signs are abraded and come closer to being mere and useless things. Analog structure, moreover, is as viscous as molasses and as difficult to manipulate.

Technological information, to the contrary, is a marvel of permanence, perspicuity, and pliability. All the same, it seems opaque and mysterious to most people. In part this is so simply because the enormous and growing complexity and speed of technological information are swamping its rigorous perspicuity. To make sense of a computer, one has to gather and conceal the electron flows of transistors and resistors in the various boxes of logic gates; collect the logic gates into adders, adders and such into an arithmetic and logic box, boxes like these into the central processing unit, the CPU and other chips into a board; and, finally, put the boards, peripherals, and a power supply in the outermost black box—the computer. To this hierarchy of boxes, extending from the transistor to the computer, there roughly corresponds a hierarchy of languages. The one that uses the vocabulary of bits and the grammar of Boolean algebra is closest to the physical structure of the machine and is called machine language. Pieces and processes of the machine language are gathered into the terms and operations of the assembly language whereon is layered the compiler language, the programming language, and so on. Just as for the common user the physical structures of the computer have coalesced and become opaque in a box, so the languages of the computer have congealed into the lingua franca of point and click.

Yet in the hands of experts, technological information promises to render reality not just perspicuous or surveyable but altogether transparent. Transparency seems to be the perfection of information about reality. Natural information makes the world perspicuous. It opens reality up, but it does not encompass it. The world remains endlessly open and eminently contingent. The signs of natural information, moreover, are transient. They come forward and recede and are unavailable to record and transport information.

Memory, therefore, was the space where information needed to be stored, and as long as writing surfaces remained precious and particularly, as Frances Yates has pointed out, "in the ages before printing a trained memory was vitally important."[4] To make it a capacious and reliable container, the Greeks, the Romans, the medievals, and even the early moderns imposed a spatial structure on memory. They learned to remember first a quiet building of many wings, rooms, and niches where then they could deposit in order what needed to be remembered.[5]

But even when the capacity to contain information was highly trained, the sense of the endlessness of reality remained acute. The well-articulated memory space could hold much information, but it was normally intended to be used over and over again for new and different kinds of contents. In antiquity, at any rate, memory was not meant to picture the entire world.

To obtain and maintain information about the shape of the world as a whole one needs to proceed from the internal space of memory to an external space of signs. Maps are the instruments that render reality not just perspicuous but surveyable from end to end. Global exploring, surveying, and mapmaking had their origins in sixteenth-century Europe. From there they spread in extension and intensity and reached what today is western Montana early in 1805 when Meriwether Lewis made a map of what, with William Clark and his companions, he had explored so far traveling up the Missouri, and of what he had heard about and expected to find on his way to the Columbia and down to the Pacific. On 7 April, he sent the map to President Jefferson.[6] Like most travelers in unknown territory, Lewis was too optimistic about how close he was to his destination. He had the Continental Divide, that seemed within reach, some six longitudes further west than in fact it is, and the area that corresponds to today's Rattlesnake Valley appears on his map indistinctly and also too far west. Once at the Pacific coast, however, Clark on 11 February 1806 completed an admirably precise and detailed map that pictures the Rattlesnake area with fair accuracy.[7] Lewis on his way back in fact crossed and recorded Rattlesnake Creek.[8]

Modern mapmaking came to this region in the twenties of this century. In 1927 the U.S. Geological Survey published its contour maps of the area. They are marvels of concentrated information and masterpieces of graphic design even though they do not reach the artistry of the delicately textured, carefully shaded, and subtly colored Swiss *Landeskarten*.[9] Mapmaking, or cartography, used to be an art and a craft. Maps were traditionally drawn on drafting tables with pens, rulers, T-squares, compasses, lettering templates, and more. Creativity, a good eye, and a steady hand were required to craft a map. Once completed, a fine map used to rival a work of art in depth of cultural information.

Though traditional maps could encompass a region, the globe, and even the universe, they failed to penetrate and dominate reality. Information had to be wrested laboriously from heaven and earth, and once committed to paper and constituting a map, the information presented a rigid and limited aspect of reality. Before Meriwether Lewis set out on his journey up the Missouri, Thomas Jefferson instructed him to collect the information needed for a map by imposing on the landmarks of natural information the great grid of cultural information that was fixed by celestial observations: "Beginning at the mouth of the Missouri, you will take observations of latitude & longitude, at all remarkeable points on the river, & especially at the mouths of rivers, at rapids, at islands, & other places & objects distinguished by such natural marks & characters of a durable kind, as that they may with certainty be recognised hereafter."[10] Jefferson was stern in his admonition on the accuracy of observation: "Your observations are to be taken with great pains & accuracy, to be entered distinctly & intelligibly for others as well as yourself, to comprehend all the elements necessary, with the aid of the usual tables, to fix the latitude and longitude of the places at which they were taken, and are to be rendered to the war-office, for the purpose of having the calculations made concurrently by proper persons within the U.S."[11]

Lewis established latitudes by using an octant or sextant to measure the angle of the sun at its highest point and using a table that for the date of the measurement listed the latitude. For the longitude he used the lunar distance method since his chronometer, though most expensive, was far from being as reliable as Harrison's masterpieces.[12] To take readings, Lewis and Clark frequently had to stay up until midnight, Lewis at several intervals measuring the angle between the moon and a prominent star, Clark recording the numbers that Lewis called out.[13] They had neither the time, nor the skill, nor even the ponderous astronomical tables required for the computation of longitudes, yet they diligently collected the raw data that in the end turned out to be inaccurate and were never used. The fruits of their nocturnal labors came to nothing.[14]

The maps that Lewis and Clark drew on the basis of what data they could use conveyed little information compared with what is found on a Forest Service map, for example. But there are unsurpassable limits to what any map can contain. If it shows a large area, the information per square mile is sparse. If it shows a square mile as measuring two and a half inches or so to the side, the map fails to provide a sweeping view. And even on a large-scale map there is a limit to how much information can be crowded onto a square inch even though the human eye is able to discriminate marks as small as 0.1 mm and thousands of different colors.[15] To collect and store large amounts of information, maps of different scales and contents are needed. But then the sum of information we have about reality amounts to an unwieldy pile of paper.

Transparency

The genius of information technology consists in making information pliable by digitizing it, making it abundantly available by collect-

CHAPTER THIRTEEN

ing and storing astronomical amounts of it, and putting it at our disposal through powerful processing and display devices.

When it comes to information about reality, geographical information systems (GIS) are the paradigm of technological information. GIS had its start in the late sixties when cartographers began to realize that the information on traditional maps had the inertia and viscosity of all analog or continuous information, and they began, in the teeth of still rudimentary information technology, to convert cartographic information into crisp and easily manipulated bits.[16] At length, they perfected the grid that at the beginning of the modern era had helped to make the world surveyable, and they set out to render reality transparent. One way of doing this is to employ a raster system where a traditional map is divided into a mass of tiny cells or pixels (one cell corresponding to as little as a 30 foot square of the real world) and the particular color of each cell is recorded in an orderly sequence. Such a procedure captures, by brute digital force, all the information of the map down to the resolution of a single cell. The alternative method, the vector system, is philosophically more pleasing. It assigns to every feature on the map a sequence of coordinate values, that is, of all the x's and y's that compose a point, a line, or an area.[17] The vector system, like writing and the electron, is both a model and a vehicle of transparency as regards the structure of information. Incorporating Euclidean and Cartesian geometry, it discloses "the form of the world."[18] Correlating locations with attributes or properties, it resembles a structure, a set of objects lawfully related.[19] In its ability to display separately the layers of rivers, of vegetation, of roads, of boundaries, of residences, or of whatever else, the vector system embodies the rise of information, the emergence of discrete signs from the dense background of things, the message that issues from the eloquence of reality.[20]

What gave GIS its sweepingly penetrating power was not so much the digitizing of traditional information as the collection of masses of new and novel data through satellites that capture and relay the information contained in the ways the radiation of the sun or the radiation emitted by satellites is absorbed, scattered, and reflected. The area around the Rattlesnake Valley that Clark had roughly sketched early in 1806 and Lewis explored on July fourth of the same year was recorded by the Landsat-5 Thematic Mapper on

20 July 1991 from a distance of 500 miles in outer space. The result can be seen on the World Wide Web by anyone, anywhere, anytime.[21] It is but a tiny fragment of the information that has come down to us by remote sensing. Beginning in 1998 the flow of information from outer space was to turn into a torrent when a new era of remote sensing will dawn with EOS, the Earth Observing System. The first generation satellite, EOS-A, alone will generate more than eight trillion bits of information a day on everything from atmospheric temperature and water vapor to surface cover and canopy chemistry.[22]

Technological information can reveal otherwise invisible things not only on and above the earth, but beneath the earth as well. The very rock and soil of a stretch of north-central Montana has been made transparent by a computer model of the geology under the Rabbit Hills oil field. The model represents an "integration of 3-dimensional seismic data, geologic and engineering models that accommodate variations of physical scale and relative emphasis of data, and finally a visualization of the integrated model set" (see fig. 18). The model consists of six or so colored and layered surfaces that represent the strata of this area, and it allows you to fly through this layered space to reach different viewpoints and perspectives on where, for example, the shafts of the oil wells penetrate and terminate in the layers of sand and rock.

Reality includes the social as well as the physical. The Census Bureau in fact was one of the pioneering agencies in the development of GIS. By displaying the ties of its data to locations, it enables you to see on a screen the population density of a town or a region. On the World Wide Web you can call up a picture that shows how the region where Lewis had encountered not a single person has filled up with people, densely so where Lewis crossed Rattlesnake Creek near its confluence with the Clark Fork and more thinly as you move up the valley.[23]

And finally, the areas of physical concreteness and social actuality are not the only ones where turbidity of information limits our view. The realms of scientific abstraction and mathematical possibility can be clouded by a surfeit of data or complexity of structure. Visualization is the device that renders such problems transparent.

CHAPTER THIRTEEN

FIGURE 18   The geological structure of the Rabbit Hills oil field in north-central Montana, from *DOE Petroleum Reservoir Characterization Project* (Online: Montana Organization for Research in Energy). Available: http://ftp.cs.umt.edu:80/DOE/home.html. 10 June 1998

Experiments with plasma, a special sort of gas that constitutes "a fourth state of matter," can yield results that are tedious to evaluate and hard to judge by conventional methods. But when digitized and graphically represented by a computer, the experimental data are available quickly and intuitively.[24] Fluid flow, this ubiquitous, intriguing, and math-defying phenomenon, can at least to some extent be made tractable and intelligible through "graphics workstation hardware and computer graphics software. Pictures of breathtaking realism and photographic detail are now within reach of individuals and certainly within reach of small companies and academic institutions."[25]

In mathematics, technological information has led to the solution of heretofore recalcitrant problems, among them the four-color problem that poses the question whether four colors suffice to color a geographical map so that adjacent regions are always distinguished by different colors.[26] At the frontiers of mathematics, computer graphics enables researchers to visualize and explore new kinds of surfaces.[27] And even neophytes are aided in their grasp of computer equations that move them to ask: What does it all mean? What is the point? What happens if you change this? Or that? Information technology can picture an equation on a screen as a curve or a surface or a sculpture, and as you change the values or a parameter of the equation, the shape of the figure on the screen will flatten, or swell, or break up. Playing with an equation or function this way gives you a feel for its power and significance.

Its power has engendered the hope that information technology will not merely overwhelm the contingency of reality but reveal its very secret. In cellular automata, the screen is divided into cells and the computer is programmed to make the array of cells pulse through state after state, each transition consisting of color changes in the cells and each cell changing according to its own state and the states or colors of its immediate neighbors. Extremely simple if well-chosen rules of change and suitably chosen initial states of the cells produce unimagined patterns of flowing, expanding, dwindling, or oscillating and more limited shapes of gliding, spewing, or devouring creatures.

The best known of these automata is called Life, and the name suggests that the seemingly unpredictable developments of organisms and populations can be modeled on a computer.[28] There is a certain loss of transparency when a computer program takes on a life of its own and unfolds in surprising and sometimes enchanting ways. Information is no longer explicitly encoded in the lines of a program but rather implied and embodied in the scattering of cells or the connections and weights of a neural network. Information takes a small step back from the explicitness of writing to the implicitness of a natural sign.

Though some mathematicians deplore the loss of intelligibility that is incurred when proofs are left to computers or are entirely sacrificed to experimental procedures, workers in the creation of artificial life seem pleased with the division of authority and creativity

CHAPTER THIRTEEN

that allows them to define the starting position and lets artificial life take its own course from there.[29] In some people such work engenders nothing less than intimations of divine creativity.[30]

Transparency of information would approach perfection if all information about reality could be united in one well-ordered information space, realizing electronically the Memory Theater that Giulio Camillo conceived in the sixteenth century. His idea was to order all knowledge in a carefully arranged fan shape. "He pretends," a contemporary observer noted, "that all things that the human mind can conceive and which we cannot see with the corporeal eye, after being collected together by diligent meditation may be expressed by certain corporeal signs in such a way that the beholder may at once perceive with his eyes everything that is otherwise hidden in the depths of the human mind. And it is because of this corporeal looking that he calls it a theatre."[31] Today's prototype of such a space, the Internet (or, more vaguely, cyberspace) is far from all inclusive and well structured. But Camillo's encyclopedic ambition survives in today's notions of spatial navigation, hyperlinking, and search engines.[32]

The Shadows of Transparency

Transparency, however, is anything but transparent and casts its own shadows of enigma and confusion. Edward Tufte, who has worked hard to promote high standards for the graphic visualization of information, concludes that principles of good design "are not logically or mathematically certain" and that most of them "should be greeted with some skepticism."[33] In GIS the challenge of clarity and economy is even more daunting, and what excellence of graphical design there is has largely failed to migrate into GIS.[34] Yet particularly in GIS, human ingenuity is indispensable if information is to be valuable. The mass of data by itself that comes from a LANDSAT satellite tends to be fuzzy and unhelpful. Firming up boundaries and highlighting differences requires a delicate interplay of computerized methods and human judgment.[35] Errors and misleading precision easily insinuate themselves into GIS maps and models.[36] There is no hope of mechanizing and generalizing the process from raw data to visualization, nor can the dream of smooth and universal information navigation ever be realized.[37]

Nor will it come to pass that neural networks, cellular automata, and the like will give us the key to the prediction of tornadoes, the cure of cancer, the control of the economy, or the simulation of human intelligence.[38] All of these enterprises have followed the path that Hubert Dreyfus described so early and well.[39] After initial excitement and striking beginnings, there are creditable successes that fail, however, to rise steadily. In time the achievements level out, and the hopes of revolutionary discoveries die a long, hard death.

This dimming of transparency occurs at the operational level of technological information where scientists use high-level computer languages, computer graphics, and software packages. But even where clarity reigns at this stratum, it rests on a substratum of machinery that is becoming concealed from the understanding of those who operate on its surface.[40] The blackboxing that is the consequence of progress in information technology encloses ever larger spaces of hardware and software. It is an unavoidable development. The larger black boxes support more powerful tools. Debilitating tedium would be the price of keeping scientists engaged with the deeper layers of machinery.[41] Yet such power can inspire a deceptive sense of ease and expertise. An equation that is quickly grasped as an interactive colored graphic is also weakly appropriated and soon forgotten. A map that is rapidly assembled through a sequence of points and clicks is far less deeply understood and less thoughtfully crafted than one that is laboriously drawn on a table. Sherry Turkle quotes this lament of an MIT physicist: "My students know more and more about computer reality, but less and less about the real world. And they no longer even really know about computer reality, because the simulations have become so complex that people don't build them anymore. They just buy them and can't get beneath the surface. If the assumptions behind some simulation were flawed, my students wouldn't even know where or how to look for the problem. So I'm afraid that where we are going here is towards *Physics: The Movie.*"[42]

Perhaps the deepest ambiguity of transparency lies in its corrosive aimlessness. The word *transparency,* like *clarity,* has a double meaning. It denotes both absence and presence. We call information transparent when the fog between us and our object of inquiry has been removed and the medium of transmission has become pellucid. But we also call clear or transparent what has become present once the fog has lifted, the objects or structures we are curious about. Transparency as a norm of clarity and presentation, however, has no intrinsic points of rest or satisfaction. On a bad day in winter the air in the Missoula valley is anything but transparent and dims the outlines of the hills and mountains. Air pollution impairs visibility. But the level and extent of pollution are themselves unclear until you climb one of the nearby mountains and see how the brownish smog fills the valley and is lapping up the hillsides as Lake Missoula did in the last ice age. Thus from high above something can present itself clearly that from the valley bottom is an obstacle to clarity.

If you imagine yourself in control of a perfect GIS, nothing any longer presents itself with any authority. Anything might as well be an impediment to inquiry. Pollution obscures the vegetation, vegetation hides the soils, the soils conceal the geology, the geology obstructs a view of the magnetic field. And all the information about the physical makeup of the globe may be thought of as getting in the way of social reality, the latter as crowding out information about the arts and sciences, and so on.

Transparency, Control, Prosperity

To be sure, this ambiguity of transparency lies, uneasily perhaps, in the farther recesses of contemporary culture. The instability of transparency is held in check where cognitive and cultural interests are well focused; and when they are, the control that is the benefit of transparency comes into its own. Thus the cognitive concern is fairly clear and salutary when it comes to the prosperity of the environment.[43] The cultural concern that rules our society is likewise clear if less salubrious. It is the craving for the unencumbered enjoyment of all the riches the world and imagination can offer.[44]

Here the path of information technology runs from the transparency of information to the control of information and from there to the control of reality. Control and the consequent sophistication of technological devices constitute the type of prosperity that is characteristic of the postmodern era. In the modern era, prosperity followed from the massive reshaping of reality and the mass production of tangible goods.[45] But by now the characteristic appliances and utilities of the technological culture have been in place for many decades. "Daily life," Bill McKibben observes, "*has scarcely changed* between 1960, when I was born, and the present."[46] What has changed, he notes, is the refinement and variety of the commodities we consume and the devices we employ.

To all outward appearances, the Boeing 707 of a generation ago hardly differs from the Boeing 777 of today. But there is a huge difference in construction. The design of the 777 was in fact a benchmark in the progress of control via technological information. Just as drafting tables were replaced by workstations in cartography, so they were in the design of the 777. It was the first paperless major transport design.[47] When billions are spent on research and development by five thousand engineers to design something so large and complex as a jet plane, it is inevitable that parts fail to connect or vie for one and the same spot in space. A sequence of mock-ups is used to reconcile design with reality. Or so it used to be when planes were designed on paper. Boeing instead used a computer model that gave everyone the same three-dimensional model to work with.[48] "The parts snap together like Lego blocks," said one of the workers on the production line where information becomes reality.[49] Compared with conventional signs on paper, the computer model eliminated more than 50 percent of what are called "errors, changes, and reworks."[50] The result was a plane that is much safer, quieter, more economical, roomier, and more comfortable than its ancestor, the 707. Much the same can be said of our houses, cars, stereos, refrigerators, stoves, and dishwashers. In all these cases, it is the insertion of technological information that produces such sophistication.

If Rip Van Winkle had slept through the past thirty years, the tangible devices and spaces of today would for the most part seem familiar to him. But one realm that has newly opened up might quite surprise if not unsettle him—the intangible region we call cyberspace.

Chapter Fourteen

Virtuality and Ambiguity

The Resolution of Information

Though the nature of information fails to warrant a rigorous intrinsic linkage of signs and things, the technology of information can produce a kind of equivalence between signs and what they are about, that is, between the containers and the content of information. When Claude Shannon wrote his seminal article on information theory, he was concerned to keep the problem he had set himself crisp and clear. So he restricted himself to the structure of signs and explicitly disregarded the question of what those signs might be about. He was interested in signals, not in the messages they were intended to convey. "These semantic aspects of communication," Shannon said, "are irrelevant to the engineering problem."[1] To support and illustrate the point, Warren Weaver has us consider a signal of one bit structure, a one, that is used to convey "the King James Version of the Bible" (a zero conveying "Yes").[2] It is hard to imagine a real world setting for Weaver's particular illustration of the signal-message distinction. But his example has the virtue of suggesting an obvious connection between the two. At least one of the two possible messages, the King James translation of the Bible, is structured as well as the signal that conveys it. The signal has a structure of one bit. How many bits does the message, the Bible have? Something like 36 million bits (or four and a half megabytes).[3] Hence the ratio of signal to message is one to 36 million. We can call this the resolution of the signal, and it is evidently very low—0.000000028 to be exact.

Though the concept of resolution is accurate and helpful in well-defined technical circumstances like those of remote sensing, it is misleadingly precise when applied to Weaver's illustration and, to make things worse, derived from an entirely artificial case.[4] But the concept of resolution has a convenient everyday meaning that makes intuitive sense and at the same time is within hailing distance of structural considerations. Thus everyone understands that present television screens will in time be replaced by better ones, the high definition television screens; and "better" means that the latter will not only be larger but also have higher resolution. Similarly a version of Bach's St. Matthew Passion recorded in the early 1940s on four-minute shellac records, retrieved by steel needles and amplified by tubes, has a lower resolution than one recorded half a century later, deposited on hour-long compact discs, read by a laser beam, and amplified by transistors.[5]

Resolution in this informal sense is a helpful guide to the distinctiveness of technological information. To use an illustration somewhat more plausible than Weaver's, imagine your daughter is visiting from Boston early in July. She tells you she has recently been drawn to the music at Emmanuel Church, and she was particularly moved by the performance she heard just before she left. "What did they play?" you ask. "Bach's Cantata no. 10," she answers.

These four words constitute a nearly natural sign, and like the medicine tree and its cultural siblings, the cairn and the altar, it carries an extensive and elaborate message, and, as in the ancestral setting, human intelligence must be strong to right the balance between signal and message if there is to be real information.[6] By the standards of E. D. Hirsch, a person of normal cultural literacy can be expected to know what "Bach" or "Cantata" refers to; but "Cantata no. 10" will mean nothing. There are, of course, degrees of intelligence here. At the top of the scale, where one would get something that approaches the full message of "Bach's Cantata no. 10," we would find the intelligence of a conductor like David Hoose or of a scholar like Alfred Dürr. Given the imbalance between signal and message, the resolution of "Bach's Cantata no. 10" is certainly low. How low is hard to say. We can determine the number of bits that make up the structure of the four words had they been written rather than spoken. It would come to 176 bits.[7] But how to number the bits in a performance of the cantata?

Now imagine a different scenario where your daughter replies to your question by saying, "Bach's Cantata no. 10. In fact I made a copy of the score. Here it is." Let us take the score to be part of the signal she gives. The message is the same as before, but obviously the structure of the sign is much more detailed. Let us assume it contains 164,000 bits.[8] Hence its resolution is higher than that of the first reply by a factor of 940. As resolution rises, the demands on intelligence decline. The score in hand, you no longer need to have the cantata committed to memory to get the message substantially. Even the musically illiterate can see from the score what instruments and voices are called for, how the text reads, and roughly how long the cantata takes. To get the entire message, one still needs a formidable musical intelligence—the skill to read music the way a literate person can read a text, and the ability to detail in imagination what the score leaves blank—the pitch, the tempo, the phrasing, and much more.

But what if a sign's structure were so elaborate that every such detail would be specified? How many bits would it take? To get an answer imagine finally that your daughter replies to your inquiry, "Bach's Cantata no. 10. In fact I bought the CD it's on. Here it is." And let the signal she gives you include the portion of the disc that contains Cantata no. 10. On a conventional CD it takes about 1.2 billion bits (150 megabytes) to record the roughly twenty minutes of music that realize Cantata no. 10. The resolution of your daughter's third answer compared with that of the first has risen by a factor of roughly seven million. One might well assume that the resolution has reached one—the structure of the sign is as detailed as the structure of the thing the sign refers to. So in one sense we have realized the ancient dream of discovering the structure that a sign and a thing, a piece of information and a region of reality, have in common. The sign equals the thing, information has virtually become reality. Contemporary culture surely endorses this conclusion. Now that, say, volume seventeen of Helmuth Rilling's Cantata series is in your possession, you no longer say that you have information about or for Cantata no. 10; you *have* the cantata; what your CD player, amplifier, and speakers produce is not something that is about or for Bach's music.[9] It is the music itself.

In proceeding from the title via the score to the disc of Cantata no. 10, we raise the resolution of a sign. There are more and more bits. But what we get is not merely more and more information, but different kinds of information. The first answer to "What did they play?" is more or less natural information—information about reality. The second answer can perform the function of natural information and tell you about some past and distant event. But the distinction of a score is that it provides information for performing the cantata. It is cultural information. The compact disc, finally, can be information about and for reality. But the technological information it contains is distinctively information as reality. Information gets more and more detached from reality and in the end is offered as something that rivals and replaces reality.

The information on a compact disc is not readable by a human being. It is microscopically condensed to begin with. But assume it is magnified and converted into a sequence of ones and zeros. The next obstacle would then be the challenge of dividing the continuous line of ones and zeros into groups of sixteen digits, for information on a CD does not consist of a single number of more than a million characters but is composed of a sequence of sixteen-digit numbers, somewhat confusingly called "words." Unlike a computer, a human being would find it impossible to divide the continuous stream of ones and zeros into the proper words. Assume then that the sequence of digits is printed with the required gaps to mark the beginnings and ends of words. Yet a human being is unable to distinguish precisely sixteen-character strings from one another if the strings are composed of only two different elements. Somehow we would have to find for each of the 65,536 possible sixteen-bit words a distinct and distinguishable symbol. That would still leave us with the task of grasping 48,000 such word symbols per second of music.

In case this was not obvious, technological information in any rendition of its basic structure is humanly unrealizable. But neither need it be realized by humans. Technological information is self-realizing if we include in it not only the pure structure of information but also the input, processing, and output devices in which it is embedded, in this instance the CD player, amplifier, and speakers. Evidently, since technological information realizes itself, the demands on the realization skills of people decline to nothing. Semantic resolution of information and human skills of realization are inversely proportional. This relation is not the inevitable result of the way information technology has developed and human beings are able to respond to it. In fact, when information technology is used for research or design, the demands on human skills can be very high. But overall, and emphatically so in the realm of leisure and consumption, technology in the narrow engineering sense and technology in the broad cultural sense have converged to obviate powerful skills and habits of realizing information. Engineering technology has increased our control over information to the point where information has assumed a distinctly new and powerful shape. Technology as a way of taking up with reality has put the power of technological information in the service of radical disburdenment. At the limit, virtual reality takes up with the contingency of the world by avoiding it altogether. The computer, when it harbors virtual reality, is no longer a machine that helps us to cope with the world by making a beneficial difference in reality; it makes all the difference and liberates us from actual reality. Of the five terms of information where, INTELLIGENCE provided, a PERSON is informed by a SIGN about some THING in a certain CONTEXT, intelligence, things, and context evaporate and leave a person with self-sufficient and peculiarly ambiguous signs.

Virtual Reality

The terms and techniques that have been developed by information theorists like Shannon remain important for the basic structure of technological information where engineers are concerned to pack as much information as possible on discs or to push a maximum of information through wires or segments of the electromagnetic spectrum of radiation (quaintly called the "ether" or the "air waves"). But where information becomes virtual reality, new dimensions need to be traced, and new terms have to be found.[10] Jonathan Steuer has distinguished between the force virtual reality exerts on a person and the force a person exerts on virtual reality. The former he has called vividness, the latter interactivity.[11] Vividness, Steuer says, has two dimensions, depth and breadth. Depth is resolution, and it is a standard that CD music meets well though it could be more finely grained and will, to the ears of vinyl disc lovers, never be smooth and subtle enough. Breadth is the number of human senses a virtual reality addresses, and in that sense CD music is narrow. Conventionally, it speaks to our ears only. But there are discs that address the eyes too and present the musicians as well as the score and in principle whatever your heart desires.

As for interactivity, a CD, compared with a vinyl disc, allows for a high degree. A CD player makes it easy to pick out a passage, to color the sound, to manipulate the sequence of selections, and more. Interactivity, as Steuer defines it, has the dimensions of speed, range, and mapping. A CD player does well when it comes to speed. It replays a passage more swiftly than a live orchestra could. But its range is limited. It does not respond to exclamations or gestures. Mapping is the dimension of naturalness and predictability. Turning knobs and pushing buttons have become second nature to us, but they are not truly natural the way speech or gestures are.

When we hear of virtual reality, we think of something more inclusive and exotic than a CD player—some device, for example, that procures the experience of flying at will through exciting and limitless spaces. A flight simulator can do this in one way, a helmet with earphones and a visual display combined with data gloves and a bodysuit does it in another. The technology that undergirds virtual reality is evolving and variable. Whatever the changes, technological information will be the heart and soul of virtual reality.[12]

Helpful as Steuer's distinctions are, they need to be complemented by terms that divide virtual from actual reality. Taking vividness and interactivity to their extremes does not lead us to the heart of virtual but rather back to actual reality. Nothing has as much breadth and depth and nothing invites as much engagement as the actual world. To distinguish the virtual from the actual, one's first inclination is to use simulation as the dividing line. "Artificial reality," an early synonym of virtual reality, suggests this kind of boundary. But the other synonym of virtual reality, namely, "hyperreality," implies that virtual reality, rather than being second best and a mere substitute, is superior to actual reality. That implication appears to be borne out by CD music. It tends to have a preternatural purity and perfection that make any live performance sound rough and flawed. Similarly, the space we traverse in virtual flight promises to have more captivating shapes and more saturated colors than anything in the actual world. Virtual reality is expected to have an overpowering brilliance that led Tony Verna to worry: "There probably will have to be a computer cutoff point to prevent the emotional sensations from getting too intense, especially for sex scenes or if you're watching a car race when the race-car driver crashes."[13]

Virtual reality is hailed as superior not only in its impact but in its variety and availability too. Even now we can easily have whatever music wherever and whenever. To race a car or ski a slope you only need to walk into the next video game parlor. Vividness and interactivity of video games leave much to be desired, but then no one has been carried from a virtual race course or ski hill with burns or fractures. Supernatural brilliance, limitless variety, and unreal availability constitute the normative identity and charm of virtual reality. The actual world seems drab, poor, and hard in comparison. But such glamour could not coexist with the gravity and duress of actual reality if the former were not discontinuous with the latter. In its pure form, virtual reality is separated from the ordinary world by a threshold that can be crossed easily and at any time and yet marks the entry into a separate reality. You have to climb into a simulator or don helmet, suit, and gloves to have a vivid and interactive experience of Superman's kind of travel.

Virtual Ambiguity

The divergence of actual and virtual reality has introduced a new kind of ambiguity into contemporary culture. Traditionally, ambiguity has implied a corrective norm—clarity. Ambiguity has been an indication of imperfection. The symbolic ambiguity of texts, scores, or plans is resolved through realization, through an enlargement or enrichment of reality that is instructed by cultural information. Real ambiguity is resolved through engagement with an existing reality, with the wilderness we are disagreed about, the urban life we are unsure of, or the people we do not understand. In either case, the resolution of ambiguity leads to clarity—the splendor of reality.

In virtual reality too, resolution is high and engagement intense. Vividness and interactivity are the terms of art that define these features. But it is characteristic of virtual reality that as resolution and engagement grow, so does ambiguity. That detachment from reality and ambiguity of information must rise together is clear from the technical sense of ambiguity in information theory. Consider for illustration the Community Hospital in Bullfrog, Montana, where during the day four physicians are in attendance whose names happen to be Alfred, Alice, Alphonso, and Alexandra. One of them is as likely as another to be called to the emergency room by a jovial nurse who likes to tease his doctors and call them by nicknames and abbreviations, "Paging Doctor Fred," "Paging Doctor Sandy," and so on. One day the nurse needs to call Doctor Alphonso and says on the public address system: "Paging Doctor Alfie, paging Doctor Alfie." Whatever the intention, the signal is ambiguous. It could refer to Alfred as well as Alphonso and is only half as informative as, say, "Paging Doctor Fonz." Failing to get a prompt response, the nurse gets urgent and shouts, "Paging Doctor Al, paging Doctor Al." Within the context of paging a physician at the hospital, this signal is entirely ambiguous and carries no information at all. In an unambiguous signal, there is a firm tie between information and reality. "Paging Doctor Alphonso" makes for a tight connection between the emergency room and the physician who is needed there. In "Paging Doctor Alfie" the bond has been loosened. It still conveys some information: "The physician who is needed is not Alice or Alexandra; it is either Alfred or Alphonso." In "Paging Doctor Al" the connection, under the particular circumstances, has snapped. There is no information.[14]

Similarly, virtual reality provides no information about the world out there and is in this regard totally ambiguous. At the same time it is or aspires to be richly and engagingly informative within. The characteristic ambiguity of virtual reality reflects the amalgamation of the sense of wealth that results from the resolution of symbolic and real ambiguity with the sense of unencumbered freedom that registers the disburdenment from reality. We can call it virtual ambiguity.

The virtual elation that is the companion of virtual ambiguity obviously contrasts with our experience of reality. Unfettered freedom has always been accessible to human beings in imagination. But flights of fancy have low resolution and little bodily engagement compared with virtual reality. Discontinuous regions of reality too have been created long ago. The builders of baroque and rococo churches had ceilings open up onto celestial spaces and sculptures suffused with supernatural light. Yet churches and theaters had unequivocal and even prominent moorings in actual space, and they would command attention rather than invite manipulation.[15] Thus both fantasy and spectacle used to defer to the authority of the real world.

So does virtual reality when it is unambiguously used for instruction or experimentation. In its characteristic sphere and glamour, however, it constitutes a separate reality. But when it does, is its ambiguity necessarily a mark of imperfection? Could not virtual ambiguity represent a positive norm?

The disputes about these questions would have little to go on if they centered on virtual realities properly so called. Even with the astounding advances in hardware and software over the past half century, truly hyperreal vividness and interactivity are still beyond the reach of information technology, and the best versions of virtual reality are limited in scope and number. There is a more inclusive space, however, that centers on the norm of virtual reality, but in its farther reaches diminishes in vividness, interactivity, glamour, or discontinuity. We have learned to call this region cyberspace.

Virtual Lives

One's view of virtual ambiguity depends on how one sees the relation of cyberspace and actual reality. On the most radical position, cyberspace envelops or supersedes the actual world. The learned version of this view has been outlined by William Mitchell, at least on those occasions where he predicts that cyberspace (the "bitsphere" as he calls it) will "overlay and eventually succeed the agricultural and industrial landscapes that humankind has inhabited for so long."[16] And something like this inclusion is suggested by Sherry Turkle when she says of a young man who has sought solace in cyberspace: "It seems misleading to call what he does playing. He spends his time constructing a life that is more expansive than the one he lives in physical reality."[17] If the actual world is merely a subregion or substratum of cyberspace, ambiguity is resolved or has at any rate evaporated. A new and all-inclusive matrix of meaning has been born.

The general term for the particular neighborhood of cyberspace that is occupied by the more expansive life Turkle describes is a Multi-user Domain (MUD), a common type of which is called Multi-user Domain, Object Oriented (MOO). "MOO space," as David Bennahum explains, "exists only as text flowing down your computer screen."[18] It contains imaginary spaces—living rooms, terraces, gardens—stereotypically described, and it allows you to enter it in any guise or description you like. You are free to choose your sex, height, looks, talents, accomplishments, whatever, and similarly the characters you meet are for the most part stylized self-representations of ordinary people. Some of the characters are entirely artificial, creatures of (limited) artificial intelligence in textual embodiment.[19]

MUDs are marginally virtual realities. Their resolution is very low, but within their narrow bandwidth vividness and interactivity are high. Low resolution is the concession that needs to be made to the limitations of information technology. Ease of constructing photorealistic settings and depth of artificial intelligence are unavailable, the former for the time being, the latter forever.[20] Thus the virtual reality of MUDs is not only confined but also impure. It constantly needs to draw on actual persons to sustain its virtual vigor.

Yet it seems that these debts to reality are forgivable since participants in MUDs can be and mostly are entirely anonymous. Accordingly, MUDs may be thought of as having the characteristic self-containment of a virtual reality. On that view, an attractive interpretation of virtual ambiguity is available: the possibility of living in alternative worlds. Cyberspace and reality, though very different in structure and texture, are morally or existentially coordinate realms and allow one to lead a life (or perhaps lives) of multiplicity and flexibility that is unattainable in real life, or RL as it is called in MUDs.[21]

Thus a man of conventional cast in real life can explore in a MUD his feminine or homosexual side, his amorous dreams, his desire for power or acclaim and so enlarge and enrich the scope of his experience. And more than exploration is possible. A woman can shift her very center of gravity from reality to virtuality and feel most fully alive when she moves in virtual space. Virtual ambiguity seems to be a burst of fluorescence that dispels the darkness of ordinary life and reveals another more luminous reality.

Real Gravity

The veil virtual ambiguity casts around cyberspace can be more or less dense. The more permeable the veil, the more intrusive the burdens and barriers of ordinary life. To secure the charm of virtual reality at its most glamorous, the veil of virtual ambiguity must be dense and thick. Inevitably, however, such an enclosure excludes the commanding presence of reality. Hence the price of sustaining virtual ambiguity is triviality. To be sure, hermetically sealed-off regions of cyberspace can be entertaining and captivating much as games, novels, and television have been in the past. They may well be more seductive and addictive than their predecessors and so intensify the familiar moral concerns about distraction, isolation, debilitation, and indoctrination. Yet they fail to be the radically novel territory of experience that would allow people to lead newly multiple, flexible, polymorphous lives.

Within virtual reality, commanding presence takes the form of personal intelligence. The latter is borrowed from actual reality—as of now, one is inclined to add. One might consider it a mere technological imperfection that intelligence needs to be imported into virtual reality and threatens to contaminate and spoil its glamour. But any intelligence that is truly virtual and known to be ambiguous in the virtual sense ceases to be engaging. We lose interest in a creature that is sealed off from the pleasures and pains of ordinary reality. Whatever the artificially intelligent voice tells us about happiness or misery is untested, unwarranted, and merely mimicked.[22]

There is a MUD containing a creature named Julia that is truly artificial and so cleverly constructed as to tease real players and make them wonder for a while how dense the veil of virtual ambiguity is that encloses Julia.[23] Like pornography and cheap novels, an entirely artificial intelligence can engage one who surrenders to a surge of unrealizable desire. Conversely MUD personae can continue to be engaging if their real identity is known to fellow players. In either case, however, we have nothing more than technologically heightened versions of traditional cases—some Pygmalion falling in love with his Galatea, a bunch of people sitting around a table, or calling one another on the phone, or e-mailing each other. The technological transformations of traditional dreams and engagements are remarkable enough, but they pale in significance with the challenges to our sense of reality and identity that information technology is supposed to have prompted through MUDs and other virtual realities.

Reflection shows, however, that virtual ambiguity, when accepted for what it is, renders virtual reality trivial, and, when pressed for its promise of engagement, evaporates. The truly gripping stories about the virtual reality of MUDs are always accounts that chronicle not the enduring splendor of virtual ambiguity, but its painful dissolution. Of course ambiguity can be sustained if participants in a MUD engage in trivial exchanges. To lead a life of many and diverse roles is possible as long as only one of the roles truly matters. A married graduate student may live the life of a medical doctor on some MUD and pursue a courtship ending with a wedding on another. Such games are feasible as long as they are walled off from actual life and kept barren of real consequences. But if the student were to lead a really polymorphous life, he would be taken to court for practicing without a license and polygamy. The human body with all its heaviness and frailty marks the origin of the coordinate space we inhabit. Just as in taking the measure of the universe this original point of our existence is unsurpassable, so in venturing beyond reality the standpoint of our body remains the inescapable pivot.

Sooner or later, the gravity of their bodily existence pulls MUD players through the veil of virtual ambiguity into the entanglements of ordinary life. Sometimes a player is cast out of his virtual seclusion when his wife discovers his amorous multiplicity. More often, players get impatient with the vacancy of virtuality and allow themselves to be drawn into reality. They set out to meet the enchanting MUD persona face to face, most often with disappointing consequences.[24] Or they try to satisfy their hunger for reality by devouring the trace of actuality that comes with a new MUD friendship. As soon as the thrill of novelty is gone and the specter of virtual vacuity rises, they move on, endlessly reenacting their quest for real engagement.

Much of MUD life takes place in this region where the veil of virtual ambiguity has expanded into a fog that opens up onto the glamour of virtuality on the one side and the hardness of reality on the other. Much time is spent traveling back and forth between the borders of a space you may traverse but cannot settle. David Bennahum, who decided to study MOOs and enlisted the help of "Jennifer Grace, a cultural-anthropology student at Duke who studies virtual communities," well describes this twilight zone. MOO immersion, he reports, "brought me closer to Jennifer. The nerd on the phone became an alluring woman named amazin (as in amazin Grace) in Lambda. Ours became my first MOO friendship. Now I spend several nights a week with her. She has become a confidante. That's the strange thing about MOOs—they pop up like trapdoors in your life and connect you to people you would otherwise have nothing to do with. All the context that separates two people like Jennifer and me disappears, leaving an illusion of intimacy: an intimacy devoid of life's accessories, the baggage that we carry with us."[25]

Virtual Fog

This impossible union of unencumbered glamour and profound engagement must sooner or later fall apart and settle for triviality or gravity. Yet the illusory escape into cyberspace does not leave reality unharmed. At least for a time, virtuality can spread a fog of virtual confusion and blur the shape of things and events with glamour and triviality. Two of the great forces of the human condition, *eros* and *thanatos*, the erotic life and the solemnity of death, have particularly suffered glamorization and trivialization. Virtuality has extricated sex from the depths of real life and made it available as a diversion that would be harmless if it were not for the disabilities and displacements it abets in real life.[26]

As for death, Tom Mandel, the *New York Times* tells us, was "one of the first (if not the first) to share on-line, with a wide audience, his own experience of dying."[27] Actually to share a person's mortal illness is to feed, clean, and change that person, to suffer the person's bursts of anger and flights of hallucination. It is to see a person suffer deeply and decay. It is to sleep irregularly and poorly and to feel confined and at times resentful. With all that it can be an occasion of grace and gratitude. In any event it is quite different from checking your e-mail when you are good and ready, to catch up on the progress of the disease, to take in the sentiments of others, to contribute one that reads, "Oh, Tom . . . Damn, damn, damn, damn . . . (Do I get TOS [terms of service violation] for that?) Sweetie . . . I am so sorry and I am so amazed that you can just get on here and blurt it out," and then to log off and go about your daily life.[28]

In detaching facets of reality from their actual context and setting them afloat in cyberspace, information technology not only allows for trivialization and glamorization but also for the blurring of the line between fact and fiction. The looseness of a tie to reality is hard to distinguish from the lack of a tie. As it happened the report in the *Times* about Tom Mandel was not its first account of a death shared on the Internet. A year earlier, the *Times* had run a piece by Jon Katz on "The Tales They Tell in Cyber-Space" that concluded with an entry from "the electronic diary of a computer bulletin board member who died of AIDS."[29] A month later, the *Times* published a letter by Peter Cortland that was titled "Cyber-Tales: Recycled Sob Stories" and began: "As a media critic, Jon Katz should spot hokum when he sees it."[30] Whom should we trust, Jon Katz or Peter Cortland?

The ambiguity of cyberspace dissolves the contours of facts, of persons, and of places. Speculation and rumor shade over into factual claims. A shy and reticent man blossoms into an eloquent and self-revealing friend on e-mail. The workplace of a woman evaporates into the nowhere and everywhere of an e-mail address. But nobody and nothing of consequence can escape reality. The truth on whether friendly fire brought down a jetliner will finally out.[31] Ralph remains a mumbling recluse, no matter his e-mail effusions.[32] Harriet does not reside in cyberspace but is an itinerant saleswoman.[33] It takes venality or complicity on our part for persons and things to remain veiled in some shade of ambiguity. Among the antidotes to the blandishments of cyberspace are skepticism and a sense of humor.[34]

One day in a graduate seminar on information theory we discussed the lamentable pallor that virtuality has cast on reality. The next day I received a "Dear Albert" e-mail message from "William Jefferson Clinton <president@whitehouse.gov>" congratulating me on my "cutting-edge work."

Chapter Fifteen

Fragility and Noise

Physical and Social Fragility

Technological information seems uniquely robust and resonant. But closer inspection shows that its vigor and reach are threatened by fragility and noise. Its robustness technological information owes to digitality. Digital structures are more enduring than material ones because they allow for faultless copying. Thus we can easily detach the pure structure of, say, an ode from an acid-eaten page and give it a fresh and perfect realization on acid-free paper. The second rendition represents the poem as authentically as the first. But if we need to replace a sculpture on a cathedral that is exposed to acid rain, the copy will differ in subtle and crucial ways from the original.

The Roman poet Horace, contemplating his work in the decades before the common era, understood the unique durability of his digital work of art and celebrated it in the ode that begins:

> Exegi monumentum aere perennius
> Regalique situ pyramidum altius,
> Quod non imber edax, non Aquilo impotens
> Possit diruere aut innumerabilis
> Annorum series et fuga temporum.[1]

In English translation, it reads:

> I have erected a monument more durable than bronze
> And loftier than the pyramids' regal structure,
> One that no voracious rain, nor violent north wind
> Will destroy, nor the numberless sequence
> Of years and the rush of time.[2]

Horace wrote the poem sometime in the twenties B.C.E. on papyrus or a shard of clay. But the original record has not been preserved. In fact, the oldest surviving manuscript is from the ninth century. Thus there is now a break of nearly a millennium in the physical or material continuity of the Horatian text. And yet the copy of the ode that I have before me, Wickham's Oxford edition of 1877, contains the very text of the ode. In spite of the vagaries of traditions and editions, the present identity and integrity of the ode is substantially undiminished and unchallenged.

A material work of art whose fame in antiquity greatly surpassed Horace's digital piece was the statue of Athena that Phidias had fashioned for the Parthenon on the Acropolis in Athens in the late forties of the fifth century B.C.E. As with Horace's manuscript, the original no longer exists, but a dozen or so copies of it have survived.[3] Yet the latter have not preserved the undiminished identity of Phidias's Athena. It has been irretrievably lost.

One should expect technological information to be still more robust than written cultural information. Since the former is binary rather than vicenisenary, as is the latter, copying has to pay attention to the least difference only and distinguish between merely two, rather than twenty-six, different symbols. If humans or machines can distinguish and read at all, this is the very least they should be able to accomplish. Technological information, moreover, is typically written, read, and copied by unfailing and objective copyists— machines. The few errors that must have crept into our copy of Horace's ode through human inattention, fatigue, or prejudice would have been avoided by machines.

Considering the massive amounts, enormous complexity, and rapid processing of technological information, one cannot but marvel at its present rigor and precision. But it is precisely its apparent power that has induced us to overlook the remarkable and many-sided fragility of technological information. The relative complexity and inelegance of the alphabet's vicenisenary digitality is just what makes it humanly legible and writable. Twenty-six letters gather information into chunks that are sufficiently compact and distinctive for humans to grasp at a glance. Thus cultural information needs relatively little to endure—paper, pencil, a teacher, and a student. It is like a community that survives and even prospers in a natural environment.

Technological information to the contrary is like a human being in outer space, totally dependent on a carapace of supporting technological devices. These, unfortunately, are physically and socially fragile, subject to decay and obsolescence; and so therefore is the information they were meant to contain and sustain. Jeff Rothenberg has sketched the problem. "The year is 2045, and my grandchildren (as yet unborn) are exploring the attic of my house (as yet unbought). They find a letter dated 1995 and a CD-ROM. The letter says the disk contains a document that provides the key to obtaining my fortune (as yet unearned). My grandchildren are understandably excited, but they have never before seen a CD—except in old movies. Even if they can find a suitable disk drive, how will they run the software necessary to interpret what is on the disk? How can they read my obsolete digital document?"[4] Rothenberg's grandchildren can count themselves lucky if the disc is still intact since the life-span of information on an optical disc is thirty years, a mere fortieth of the age of the manuscript that contains the first extant version of Horace's ode. On a magnetic tape, information would have lasted one year, on a videotape two years perhaps, and on a magnetic disk five to ten years.[5]

If Rothenberg's grandchildren have somehow recovered the stream of ones and zeros, they then need to know how to chunk or parse it into groups that represent characters. Should they take chunks of four, eight, sixteen, or thirty-two digits? Assuming now that somehow they have determined the chunks to be eight bits long, what do the chunks represent? Letters, numbers, sounds, or images? And if they are letters, how are they structured? That depends on the particular word processing program that was contained in the computer Rothenberg used to write the document, a program which is itself a big piece of technological information needing to be parsed and interpreted.

Through the passage of time, technological information becomes physically fragile because so do the traces and the media they are inscribed on. Technological information is socially fragile because of our heedless rush toward more powerful technologies that condemn older ones to obsolescence and illegibility.[6] Social fragility is a threat to technological information not only in time and due to thoughtlessness but even now because of mischief and crime. Cultural information is by its nature scattered and refractory. It can be stolen and destroyed, but only with exertion and a piece at a time. The information in cyberspace, however, is so tightly interconnected and so quickly, if not always easily, reached, that ill will has become more potent and destructive.[7]

Structural Fragility

Yet even when discs are new, programming languages are understood, and intentions are constructive, technological information is structurally fragile. Its massive complexity is intended to deal with the contingency of reality, yet growing complexity gives contingency more and more openings for revenge.[8]

The programs that regulate telephone systems or space launches contain millions of lines. Human beings can no longer comprehend how all those lines cohere or collide. Redundancy, standardization, mathematical analysis, and other tools can reduce the number of flaws or "bugs" in a program. But at the end of the day, some bugs will escape detection and can cause communication and financial systems to break down, airplanes to crash, and patients to get killed.[9] The effect of technological information on human health and safety is still hugely beneficial. But the hope that bits, programs, and computers would gradually outgrow the need for human care and drive the recalcitrance of reality into total submission has not been fulfilled. It is only the constant and strenuous vigilance of engineers and programmers that keeps the large computerized systems alive and revives them when they have gone down.

The overt structural fragility that besets technological information due to size and complexity can be traced in part to a profound and concealed fragility of the tie between technological information and reality. The initial vision of technological information had it that, if only we connected the elementary measures of information with the fundamental bits of reality, then the grammar of technological information, Boolean algebra, would preserve meaning as technological information rose from simplicity to complexity. The right syntax would take care of semantics. Boolean algebra does directly model the way we regularly use "not," "and," "or," to modify and compound sentences. If our premises are true, then our conclusions, no matter how complex or distant from the premises, will be true as long as we observe the rules of Boolean algebra. But this is the exception. In general Boolean algebra does not represent the characteristic structures of the edifices of reality, their columns, arches, vaults, and walls. It functions more like the small scale structure of mortar that can be used to connect bricks and bits into larger components but cannot tell you how. Bricks and mortar have given rise to St. Michael's in Lüneburg that once resounded with Bach's singing. But more often, bricks and mortar have been combined into plain, ugly, or ramshackle structures.

When we hear of a computer model of this or a digital version of that, we tend to assume that such technological information somehow reveals and represents the structural essence of whatever segment of reality. But the digital rendition of, for example, a cantata mimics the appearance rather than discloses the structure of the piece. Music reaches the ear as sound waves, patterned compressions of the air. These sound waves can be captured as an undulating line on paper or in the grooves of a vinyl record. To digitize this information is simply to measure the varying distance of the wave form (its ups and downs) from some base line and to express each distance in a binary number. There are 65,536 such numbers you can choose from, and they may be thought of as the letters or notes of the technological notation for music. But just as someone who takes dictation needs to know next to nothing about syntax and semantics, so the technological notation for music involves no knowledge of musical structure.[10] Traditional notation explicitly reflects at least some of that structure—the meter, the key, the themes, the repeats, the instrumentation, and more.

There is a continuum in technological information from mimicking to disclosing the structure of reality. Models of fluid dynamics that simulate the way a plane interacts with the air have been most revealing.[11] Models of how the economy behaves have been famously fallible.[12] When it comes to social systems, weather systems, and environmental systems, there is a real question whether there is a "system" out there at all, that is, an array of significantly distinctive objects lawfully related. Regarded through the lens of computer programs and mathematical equations, much of reality will look like an impenetrable thicket of complexity and contingency. In such cases, one can still cobble symbols and rules together and build something that in some significant respect behaves like the thing one wants to model. But this is a precarious approach whose fragility is often overlooked.[13] The tenuous tie between model and reality can easily snap, and if we place our trust in it when making social or ecological decisions, we are courting disaster. Even in the supposedly successful model of fluid dynamics, "a simplifying assumption that seemed so obvious no one questioned it for 50 years," has turned out to be in error.[14]

Ironically, complexity and contingency of reality most severely enfeeble the model of reality that was to supersede reality entire— virtual reality. The curse of overly simple assumptions and impossibly strong claims haunts work in virtual reality at least as much as workaday computer modeling. Virtual experiences are powerful only to the extent that a person's complicity with the illusion compensates for the austerity of information. When it comes to virtual reality, we are still so impressed by the fact that it can be done at all that we forget to ask whether it is being done well. You may be able to fly at will through colorful expanses, but this is a silent and untouchable world of the simplest geometrical shapes. To texture real things in photorealistic detail, to add auditory and tactile displays and coordinate them exactly and on time requires computing power that vastly exceeds what is on the horizon. Compared with the vividness and interactivity of actual reality, virtual reality turns out to be a pale and brittle world and is bound to remain so.[15]

Cultural Fragility

Finally, technological information is fragile culturally in the way the life of a parasite is. Not that parasites fail to have lives and charms of their own. Mistletoe is thought to have a magic that fir trees lack. In any case, technological information draws much of its life blood from real and traditional culture. Consider the effect that the growing hardware and software power has had on the texture of information technology. Rather than leading us more deeply into a distinctively technological realm, these advances have brought the relatively telling austerity and structural intelligibility of the early operating systems back to a rich if opaque semblance of the everyday world. The information that is available to an office worker is assembled on the monitor as the picture of a desk top with a phone, a Rolodex, a postcard, a notebook, and a calendar. In the background is a clock, an inbox, an outbox, and a filing cabinet. A more inclusive picture of what information is available, Apple's e-world, looks like a little town with an info booth, a mail truck, a newsstand, a business and finance plaza, a learning center, etc.[16] To reach and handle information all one needs to know is how to point and click.

A less obvious dependence on actual things and practices is to be found in the cultural sphere of cyberspace. The actual world will always be the school of experience and the storehouse of material for more or less virtual pleasures. What animates a golfer on a virtual course and lends the hyperreal setting a modicum of charm and personality is the spirit of something like the gorse-lined, windswept links of St. Andrews' Old Course in Scotland.[17] The student of art and history who is browsing the Microsoft version of the National Gallery in London is the beneficiary of the foresight and acquisitiveness of Victorian England.[18] And seeing Holbein's *The Ambassadors* come up on the screen is a tepid encounter compared to walking through the neoclassical spaces of the museum and coming face to face with the painting that celebrated the opulence of French nobility just when the balance of power began to shift toward the English and the middle class. No amount of helpful commentary can substitute for the grandly bourgeois and British setting of Trafalgar Square whose center is marked by the monumental column that supports Lord Nelson, one of the protagonists in Britain's rise to world power. But is it not simply a matter of perfecting technological information to the point where users of the *Microsoft Art Gallery* can have an interactive full motion video that furnishes them with the experience of strolling through the museum and ambling out on Trafalgar Square to admire the Nelson Column? The highly impoverished quality of such walking aside, virtual reality, however expansive, is finally bounded and connects of itself to nothing while the actual Gallery borders on Trafalgar Square adjoining in turn St. Martin's Church and neighboring Charing Cross and so on in the inexhaustible texture of streets and focal points that is London. There is an appearance of creative energy on the Net and the Web. But it is a frail show of brilliance as long as greatness remains promissory and substance is borrowed.

The parasitic nature of technological information is more poignant in the social region of cyberspace. When David Bennahum was drawn to the semivirtual "alluring woman named amazin," did he not wish to be amazed by an actual woman, someone like Laban's daughter Rachel who "was beautiful and well favoured"?[19] But if amazin is the semivirtual product of technological information, Rachel, for all we know, may be a figment of cultural information. Why should Rachel have a greater claim on our engagement? The crucial difference between the charms of a MOO and the story of the Bible is that the former reduces persons to disposable commodities while the latter celebrates a person whose beauty was so commanding that "Jacob served seven years for Rachel; and they seemed unto him but a few days, for the love he had to her."[20]

The dependencies of technological information on nature and culture do not directly extend from St. Andrews to virtual golf or from biblical piety to an e-mail correspondent. The "woman named amazin (as in amazin Grace)" has only a tenuous link to the divine power "that saved a wretch like me," and there is surely nothing objectionable in the playful allusion that such attenuation makes possible. What is remarkable is the gradual devolution of significant structure that is illustrated by amazin and pervades technological information. One is reminded of the fate of the abbey church of Cluny, once the grandest of Romanesque churches, more than six hundred feet long and crowned with seven towers. Between 1798 and 1823, in the train of the French Revolution, the church was for the most part blown up to make room for stables and to serve as a quarry.[21] Nothing quite so brutal and thoughtless is wrought by the cyberrevolution. But there is an analogous leveling and a failure to build something that has the orienting power of the edifices we have been quarrying our materials from.

The Origin of Noise

The fragilities that beset technological information should be easy to accept. They are far from invalidating or endangering the astounding accomplishments of information theory and computer science. They do however suggest limits and caution as regards the advancement of information technology and a strategy of incorporating technological information into contemporary culture deliberately and realistically. The task, it would seem, is a matter of commensurating the fluidity of information technology with the stability of the things and practices that have served us well and we continue to depend on for our material and spiritual well-being—the grandeur of nature, the splendor of cities, competence of work, fidelity to loved ones, and devotion to art or religion. What is needed is a sense for the liabilities of technological information and an ear for the changing voices of traditional reality.

But instead of calm attention we get distracting noise. As information theory illustrates, fragility is a source of noise. When it overtakes a record, tape, or disc, there is an encroachment of crackles, hisses, and stutters on the musical information that is stored on those media. To be sure, noise can also be an inconsequential addition to the intended information.[22] When the nurse in Bullfrog's Community Hospital pages a doctor while the *Four Seasons* can be heard in the background, the music is technically speaking harmless noise. But when halfway through the announcement of a name static intrudes and crowds out the rest, as in "Paging Doctor Alf . . . (snap, crackle, pop, hiss)," the noise is harmful, having degraded the intended information by causing equivocation.[23] It is no longer clear whether Alfred or Alphonso is being called.

The fragility that is causing noise in the realignments of culture is not of the technical sort, nor is it a direct consequence of the physical, social, structural, or cultural fragilities noted above. It is rather the profound frailty of the voices of reality that has allowed so much loudness and shrillness to invade the public conversation; and, as in the technical case that illustrates the issue, cultural noise is merely annoying at certain times and truly injurious at others.

The growing silence of nature, art, and religion at the beginning of the modern era was one of the great historical contingencies that have antecedents and causes but no lawful explanations. These epochal events needed to be recognized and met by persons and peoples. Democracy and technology have been the dominant responses, and we, the heirs of these projects, owe our pioneering ancestors admiration and gratitude along with a measure of sorrow about the costs of technology and the imperfections of democracy. The present shift of technological energy from physical space to cyberspace is one indication that the modern project has reached a divide and needs to cross over into another kind of culture.

But this is a lesson we have a hard time learning and one, unfortunately, we can continue to evade indefinitely. The modern project, for all its creative energy and astounding accomplishments, has been haunted by an unshakable sense of failure, an apprehension that the fulfillment of its promise of prosperity and liberty has remained just beyond our grasp. Daily life, to be sure, has its pleasures and contentments, and public affairs can rise to moments of celebration and triumph. Yet any prospect of a technological breakthrough is eagerly embraced by the public and touted by the mavens as the dawn of a revolution that will at last usher in a life of unspeakable riches and limitless self-determination as though ever since the beginning of the modern era we had toiled in destitution and oppression.

To all appearances the fragilities of information technology are not evident enough as yet and the frailty of the voices of traditional culture is still too dispiriting to keep the noisy rhetoric of radical innovation and liberation from intruding into the more thoughtful reflections on information and reality. The most vociferous advocates of technological information describe its rise as "a revolution that makes political revolution seem like a game" or as "the most transforming technological event since the capture of fire."[24] A more moderate and specific salutation has become a widely used and nearly unchallenged formula to characterize the coming information age.

This new age, as an early instance has it, "will change forever the way an entire nation works, plays, travels, and even thinks."[25]

Education

The hyperbole of the revolutionary declarations and the blandness of the standard prognosis relegate these pronouncements to the realm of harmless if distracting propaganda. More injurious noise is spreading in higher education. Research and teaching no doubt are intimately related to information. The former produces it, the latter imparts it. Hence it is a plausible assumption that a fundamental change in information will transform education. In part, this assumption is already beyond doubt. The computer can be a uniquely powerful research tool. There are mathematical structures, natural phenomena, and social patterns that would have remained unreachable and unknowable without computers. More generally, most every scholar has benefited from the information revolution. The world of typewriters, card catalogs, and snail mail seems impossibly slow and cumbersome today. The scholar's world has become more expansive and perspicuous in the sense that more people and data are reached more rapidly and controlled more easily. Students are being trained to acquire the requisite computer skills for research, industry, and commerce.

These changes, however, leave a large part of instruction more or less in its traditional shape and fail, it seems, to follow through on the possibilities of technological information. Both natural and cultural information have sponsored characteristic ways of gathering and transmitting information. Why should not technological information follow suit?

In a world of natural information, signs are spaced and timed to refer to particular things. Research is bodily exploration and tangible experimentation. Teaching is showing and demonstrating, learning is imitating and practicing. This sort of education has migrated from its natural origins to the cultural setting under the heading of apprenticeship. A medieval youngster would be shown how to hold and wield a pickax to rough out a chunk of rock into a regular block.

Later he would be handed a mallet and a chisel and taught how to carve corbels, lintels, and capitals. At last the carver may have become a sculptor of ornaments and figures.[26] At each stage, the apprentice would be instructed to recognize the signs that disclosed the nature of things—the grain of sandstone, the cracks in a handle, the wear of a chisel, the shape of a template.

A new kind of research and teaching arose with the development of cultural information. Texts, plans, and scores allowed for the conception of doctrines, buildings, and music that were far richer and more intricate than what unaided imagination and memory could have supported. Now the mechanics of literacy had to be taught. More important and distinctive, however, the skill of realizing cultural information had to be acquired. Since most learning came to be set down in writing, reading and comprehension became focal points of education.

In antiquity and the Middle Ages, when texts were rare and precious, the sharing of texts through reading them aloud, the *lectio* in Latin, became central. As medieval instruction developed, lectors (readers) became instructors too. They would not only recite but also comment, explain, and illuminate the text. In time such reading and instruction came to be known as *lectura*, whence our word "lecture." Thus lecturers became the warrantors and anchors of a text. Through their person they would bring the meaning of a text alive in reality, that is, at a particular time and place. Lecturing, along with the institution it developed in, the university, has passed into contemporary higher education. Apprenticeship is more prevalent in vocational training and the earlier phases of education, but it continues in the collegiate environment as well in the form of labs and seminars. All these traditional forms of teaching, however, are under attack by the proponents of technological reform.

The name they have given to education via information technology varies: distance learning, distributed learning, hyperlearning, and more. Under any appellation it is the extension of the promise of liberation and enrichment into a realm where authority, the confinements of space and time, and the burdens of one's shortcomings still lie heavily on the student. Once education is moved from campuses into cyberspace, all these fetters fall away. The student becomes the sovereign who can choose the material, the method of presentation, and the time and place of studying. But learners not only change from passive recipients to active choosers of information, they are also freed from the injuries that prejudices of gender, race, class, or physical appearance might otherwise have inflicted on them.[27]

In the traditional circumstances of education, as the proponents of technological information see them, the teacher is someone like Joseph, the governor of Egypt under the Pharaoh. The students are like Joseph's brothers who come asking for provender and must accept whatever Joseph parcels out.[28] In the technological setting, students are active choosers rather than passive recipients. They are like the customers of a well-stocked twenty-four-hour supermarket who have made their lists, come in when convenient, move along the shelves, and assemble whatever they need and want. Teachers are like the butchers and bakers in the back who produce and package the merchandise, like the clerks who keep the shelves full and up-to-date, and like the store managers who are ready with help and advice.

There is of course a continuum between the traditional lecture and technologically transformed learning.[29] But its central portion is unstable and likely to disperse and move toward one or the other extreme. Thus a lecturer in a high-tech classroom may digitize her slides, diagrams, lists, and hand-outs and use the conveniences of technological information to display this material easily and impressively in the classroom and store it on a server. She may also set up a list or bulletin board on a server so that students can share opinions and assignments. Finally she can invite students to communicate with her via e-mail. Since on occasion students are unable to attend a lecture, the technologically empowered professor will provide the lecture notes for the days in question on the server. And since the requests for notes are frequent or unpredictable, she may as well store a complete set of carefully crafted notes on the computer.[30] But now, assuming that every student has a computer and is properly wired, all the material of the course as well as interaction with fellow students and the instructor are available to any student anytime. And why not anywhere? At least as far as this course is concerned, there is no need for any student to take up residence at a campus and submit to the constraints of time and place, of persons and institutions. And if this course is so available, why not all courses? The tendency that is at work here illustrates well the momentum of technological information. Digitization and electronic transmission, storage, and processing make every kind of information controllable and available. Hence it seems backward and inconsistent to leave any information shackled to the inertia of analog media, to the immobility of a definite place, or the inconvenience of a particular time.

Teaching by way of apprenticeship finds its completion in the artisan. Schooling via lecturing culminates in the scholar. What kind of expert does distance learning aim at? Since in cyberspace prodigious amounts of information are easily available, it seems foolish to commit information to memory. What the student needs are higher order skills, learning how to learn, finding whatever information is needed, and solving problems generally.[31] The goal, then, of education in cyberspace is to produce the learner, the person who has learned how to learn but otherwise knows nothing.

There is a remarkable parallel between the mind of the learner and the structure of a personal computer. The latter, in its quiescent state, knows nothing. It knows *how* to retrieve and process information, and its storehouse of information, the hard disk, can be huge and contain as much information as a small library does. But its working memory, the main memory, is empty every time the computer begins to work. In fact, even what it has permanently learned about how to learn is reduced to a minimum so that it is free to acquire different or ever more advanced systems of how to operate. In both the computer and the learner, the complement to "having the world database at your fingertips" is to have nothing in your head.[32]

The kind of individual sovereignty that information technology is to install in education has a long ancestry. To be sure, there was much oppression and mindless memorizing to rebel against in the nineteenth century when educational reform movements began. But as has happened in modern culture generally, the line between genuine liberation and indulgent disburdenment was thoughtlessly crossed. On the face of it, the idea that learning can always be a matter of joyful and spontaneous discovery is surely attractive. It gradually made its way through the educational establishment of this country and has left a deep imprint on primary and secondary education.[33] The explosion of information technology has given the traditional promise of unencumbered educational prospering renewed vigor and a novel specificity. Learner liberty and sovereignty are now taking on higher education.

There is a coterie of educators and salespeople who have made the primacy of information technology their clangorous cause. More remarkably, most academic leaders are enthusiastic about the blessings of the Internet. They tend to think of technological information as a powerful new fertilizer that will invigorate and extend higher education, root and branch.[34]

Higher education illustrates how due to the general frailty of contemporary culture and in the face of the fragility of information technology hopeful clamor gets in the way of calm reflection. Billions of dollars are dedicated to educational hardware and software, but next to nothing is spent to get reliable information on the costs and benefits of the expenditures.[35] But what little is available, when put in the larger context of information and reality, suggests that our enthusiasm for computerized learning and research will serve students and scholarship poorly.[36]

The rhetoric of recasting education within the framework of information technology is well attuned to the promise of technology and, in fact, to the implementation of that promise. The disburdenment from the constraints of time, place, and the decisions of other people is the unique accomplishment of modern technology and finds its everyday realization in consumption. Supported by the machinery of technology, consumption is the unencumbered enjoyment of whatever one pleases. The pleasures of consumption require no effort and hence no discipline. Few proponents of course would claim that distance learning will be effortless. But they fail to see that the discipline needed to sustain effort in turn needs the support of the timing, spacing, and socializing that have been part of human nature ever since it has evolved in a world of natural information. Just as important, they fail to see that it would be a Pyrrhic victory to deal with social prejudices by concealing them through a technological fix. People possessed by bias have most often learned to overcome it when they and their supposed inferiors were engaged face-to-face in a common and demanding task. Virtual colleges would help to perpetuate a problem that actual colleges have begun to solve.

The traditional campus setting of higher education recreates a good part of the spatial and temporal order of natural information. Lectures have their particular times and places. They are held in cavernous halls with most of the attention focused on one person. Seminars are conducted in small rooms with much interaction between instructor and students. Labs have a characteristic smell and engage hands as much as eyes and ears. Libraries provide places for quiet reading, coffee houses surround us with their comforting hum. Even at the small scale of the office, writing here, reaching for a folder there, getting up to take a book from the shelf—these minor bodily engagements help to space and time information and to make it more likely to become knowledge.

Technological information, to the contrary, comes endlessly and relentlessly pouring forth from one source to address an immobilized body via one sense. Or so it would if personal computers were a truly rich information source. As it is, the prohibitive imbalance between abundant technological information and severely stunted capacity is righted by reducing information to a thin trickle, tricked up into colorful bubbles.

Research, Scholarship, Fiction

Even in serious and sober research, the preternatural resonance of cyberspace invites cacophony rather than clarity. The "cold fusion" phenomenon of the 1980s was itself, as it turned out, a matter of sound and fury rather than of sound inquiry and information. The electronic newsgroups, praised for their richness and freedom of information exchange, were of little help in extracting signal from noise. Even in the strictly technical sense, according to Bruce Lewenstein's estimate, "30 percent of the volume of the news was pure noise."[37] But noise, taken more broadly and culturally, was the real bane of the discussions. The boundaries between fact and conjecture, importance and irrelevance, truth and error became fuzzy. As Lewenstein has it, "the low cost of entry may also lead to a low cost of error."[38] Noise created ambiguity, ambiguity got in the way of knowledge. Lewenstein concluded that with one exception "active researchers followed the net only rarely (perhaps a few times a year) if at all. Most likely, they found the 'signal to noise ratio' of the public newsgroups too poor for the nets to be useful."[39]

Ancient Greek texts used to be the training ground of many an eminent mind. Before one could hope to fathom the power of Homer and the wisdom of Plato in their own words, one had to memorize a sizable vocabulary and, worse, endless paradigms of declension and inflection. Translation proceeded cumbersomely by analyzing the forms of words and construing the order of sentences. Resort to translations in the vernacular was considered a felonious shortcut. Learning ancient Greek was like traveling laboriously through rocky and mountainous terrain, grammar and dictionary being one's staff and provender, until at length memory and comprehension were sufficiently strong to allow for engaging journeys of discovery. But now diligence and information technology have provided aspiring classicists with a resourceful explorer of magical gifts, an automated database called *Perseus*, that some day soon will furnish a complete morphological analysis for every word of the corpus of ancient Greek texts and for every sentence an English translation.[40] To travail and travel through a text in the traditional way must seem as sensible to today's tyro as making one's way from Montana to New York on foot.

The deceptive sense of facility that such a database inspires in the neophyte is bound to infect mature scholars as well. With the entire corpus transparently spread out before them and ubiquitously and instantly inspectable, they are in a position analogous to that of geographers in possession of GIS who no longer need to survey and record a creek on the ground when remote sensing describes its course much more quickly and accurately. Similarly the *TLG*, the *Thesaurus Linguae Graecae*, the basis and predecessor of *Perseus* that contains all ancient Greek writings, tends to disburden scholars of the "need to read and re-read the texts themselves" that are, after all, "the fundamental source of knowledge and inspiration" as Karen Ruhleder has put it.[41] Texts get flattened out, and scholars get detached from their work.[42] And while texts become superficially more available, the machinery that supports such availability becomes invisible and unintelligible.[43] *TLG* is the classicist's GIS.

What holds for cognition goes for fiction. Here too rebellion against oppression has conspired with information technology to produce sonorous promises. Reading a traditional linear text, says Jay David Bolter, "we are compelled to narrow the possibilities into a single narrative."[44] But when through digitizing and information processing a text has become a network, it no longer has a "univocal sense; it is a multiplicity without the imposition of a principle of domination."[45] It has proven difficult, however, to move with vigor and grace in the creative space information technology has opened up. Robert Coover had these observations about fiction in cyberspace: "And what of narrative flow? There is still movement, but in hyperspace's dimensionless infinity, it is more like endless *expansion;* it runs the risk of being so distended and slackly driven as to lose its centripetal force, to give way to a kind of static low-charged lyricism—that dreamy gravityless lost-in-space feeling of the early sci-fi films."[46]

Even a celebrated hypertext novel like Stuart Moulthrop's *Victory Garden* suffers from the endless, aimless echoing of cyberspace.[47] Whether one enjoys or dislikes Moulthrop's clever wordplay, knowing allusions, and saucy episodes, one is constantly haunted by alternatives rejected, connections missed, passages overlooked. There must be at least one engaging path through the labyrinth of the *Garden*. But to be told what it is would be to slide back into a conventional novel unconventionally (and uncomfortably) displayed. To design such a path through exhaustive labor is time doubtfully spent. To be suspended in pointless loitering is certainly a new kind of experience. But the praise that has been heaped on the condition is entirely parasitic on its opposition to traditional fiction.

Business

The most stringent test of the robustness and resonance of technological information has been undertaken in business. The test has been massive. Over four trillion dollars have been invested in information technology between 1960 and 1995.[48] Today about one trillion dollars a year are spent on capital improvements and maintenance.[49] The expectation has been that the explosion in our ability to control information would lead at the least to an explosive growth of productivity—the wellspring of economic prosperity—and very likely to a new era of leisure and affluence.[50] The measurement of productivity is a difficult and contentious business, and whatever the power of information technology, it has failed to make its effect on productivity entirely transparent.[51] But the information we have shows clearly enough that during the very period when investments in information technology were climbing steeply, productivity gains have flattened out.[52]

The reasons for this disappointment range from fragility to noise. Once important functions of business are entrusted to information technology, breakdowns (inevitable given the massive complexity of hardware and software) bring everyone to a helpless halt. When "the computer is down," there is no help in paper and pencil, in walking and checking, in asking and advising; the system needs to be brought up for things to get rolling again.[53] There are both bigger and smaller calamities than system breakdowns. Large and expensive investments in hardware and software resist implementation or repair and reduce productivity or have to be abandoned entirely.[54] A glitch on a particular piece of personal computer software can lead people to spend hours on a search and destroy mission, spent in grim frustration by some and with gleeful eagerness by others, but unproductively in any case.

In addition to the noise generated by failure, there is a systematically unproductive hum that tends to go with information technology. Information, made abundant and disposable by technology, can lose its bearing on reality, and signs proliferate without regard to things. Much that is monitored, collected, compared, transmitted, and displayed consumes equipment and time without promoting what needs to be done. Worse, once information is extracted from reality and located in cyberspace, it easily neighbors on news, gossip, and games, and people seated in front of their computers slide nearly undetectably from one to the other.[55]

Information technology has become part of the culture and unquestionably helpful in research, education, business, and the arts and letters. But the productive uses are beset by brittleness and clatter due to the distension of information technology. Some of the excess will correct itself and leave technological information with greater clarity. But more decisive action may be needed in addition. The inherent fluidity and facility of information technology may move us to consider a radically different way of presenting information, some method of selecting, stabilizing, and secluding information so that it will invite quiet attention, and a manner of making it spare and austere enough to engage memory and imagination. We may find a new regard for an old vessel of information—the book. And when we have recovered the book, we may want to restore the place that used to be dedicated to the quietude and concentration the book inspires—the library.[56]

Information and Reality

The Lightness of Being

The ecology of things used to enforce an economy of signs. The technology of information, however, has loosed a profusion of signs, and there is by now a rising sense of alarm about the flood of information that, instead of irrigating the culture, threatens to ravage it.[1] Technological information in the strict sense constitutes the distinctive and most energetic current of this development. But the more traditional stream of cultural signs has become roily and swollen as well. Information is about to overflow and suffocate reality.

Things will get worse before they get better. There is still a period of information expansion and integration ahead of us. What has been impounding information to some extent are the partitions between different carriers of information, of print vs. radio vs. television vs. the Internet vs. the telephone vs. video cassettes vs. compact discs vs. etc. One of the divides is structural—the division between analog and digital information. Elsewhere information flows are separated by their channels and the devices that present the information to us. Digitization will expand until it has breached the structural dams. Integration of hardware will advance as it already has in the "convergence technology" that fuses television sets with telephones and personal computers. But just as a contemporary home has innumerable clocks and timers rather than one central chronometer, so the household of the future will be saturated with all kinds of information devices. What matters is that clocks are synchronized and information appliances synchromeshed, that is, made to shake hands and converse with one another. Incompatibility will be the most enduring barrier.

Inevitably there will be some flood control too. The Internet in its presently free flowing luxuriance cannot survive. The selfless enthusiasm of hackers and the high-minded support of public institutions, so crucial to the first flowering of the Internet, will both decline. Hackers are getting tired, institutions will get stingy. Commerce will step into the breach, drain the swamps, channel the currents, erect dikes, build reservoirs, and install locks. All users will become customers and will have to pay either with money or with attention paid to commercials. Along with the commercial firming up of the information infrastructure, there will also be some beneficiating of information content. Individuals will come to realize that it is beyond their capacity to spot and evaluate information. Mechanical devices such as infobots or intelligent agents are of limited use. They can at most assure that the information with the right content or frequent readership is brought up. But they cannot guarantee that the stuff is accurate and worth viewing or reading. Thus media organizations that have earned our trust in collecting, editing, and warranting information will survive or reemerge. They will also satisfy our desire to know what everyone knows and so to have some sense of national awareness and belonging.

After a period of experimentation and dissolution, order will reassert itself in social organization as well. William Mitchell's idea that our tangible ways and places will become secondary to the orders and structures of cyberspace will be tried and tested in the developments of telecommuting to work and education.[2] No doubt there will be more flexibility to the times and places we learn and labor. But few people have the motivation or discipline to work well in solitary and seductive environments. Most of us fall victim to the combination of boredom and comfort that distinguishes today's domesticity. Once more, the requirements of human nature and the rigors of commerce will lead us to keep the tangible settings of home and work segregated so that we can continue to consume appropriately and produce efficiently.[3] Similarly, if the promoters of the virtual university are true to their promise of "competency verification," they will discover that many kinds of competence cannot be verified mechanically at all and that of those that can, few will be acquired by virtual students.[4]

CONCLUSION

Does this account of the expansion, integration, and organization of the new information culture do justice to the enormous innovations in hardware and the evolutions of software that are still to come? No doubt technologically literate observers will have occasion to be impressed, surprised, and astonished. At the same time they will have to recognize, if they are not convinced already, that universal and accurate speech recognition, complete automatic translation, deep syntactic and semantic analysis, and, a fortiori, artificial intelligence are unreachable goals for computers and will forever limit our ability to control information.

Most important, however, no matter how technically accomplished and admirable the breakthroughs in hardware and the advancements of software, the public will finally remain unimpressed. We are witnessing the ironical spectacle where on the public side of the stage futurists exhaust their powers of imagination and description to paint a captivating and amazing picture of the future, and where on the private side the ordinary Joe and Josephine sit on their couch having anticipated and discounted the greatest marvels of information technology. Regardless of how large, fine-grained, three-dimensional, and photorealistic the displays, no matter how universally accessible and smoothly integrated every imaginable piece of information, the rule of simple desire, having conceived of flying carpets, genies in bottles, and magic wands, has always and already preempted the most sophisticated feats of technology.

Thus with all its impending expansion, integration, organization, and innovation, the information revolution, if it stays on its present trajectory, will devolve into an institution as helpful and necessary as the telephone and as distracting and dispensable as television with an unhappily slippery slope between its cultural top and bottom. The characteristic mood of the information age begins to surface when the prophets of the new era try to get beyond grand generalities and reach for specificity. In the future, they tell us, we can get instant compliance to commands such as "List all the stores that carry two or more kinds of dog food and will deliver a case within sixty minutes to my home address."[5] And we will have the pleasure, before viewing a film on command, of being prompted "How about trying a delicious pizza from Marcello's?" and we will have the luxury of being able to "Click yes, and then select plain cheese, mushroom, sausage, or other toppings."[6] The crashing banality of such scenarios is matched by the generally dreary atmosphere that pervades gospels of cyberspace, be they science fiction or sober prognostication.[7]

The real effects of technological information will be subtle and light, but epochal all the same. For better or worse reality will become lighter, both more transparent and less heavy. We are still learning to see the world in the light of technological information, and as we do, expanses formerly too broad, structures once too fuzzy, matters at one time too dense are all becoming clear and bright. For humans who "by nature desire to know," this is a wonderful gain.[8] Where the luminous quality of technological information becomes specular, however, the remaining contingency of reality seems all the more dark and dense. The preternaturally bright and controllable quality of cyberspace makes real things look poor and recalcitrant in comparison. To be sure, reality at bottom remains inescapable and unfathomable. It is the ground on which the ambiguities of technological information can be resolved and its fragilities repaired. Yet on its surface reality appears lighter and often lite.

Information technology has begun to invade the plains and valleys of Montana's ranches. About half of the ranchers (often the women) use computers regularly for financial and cattle production records. They are significantly more satisfied with their performance of these chores than their paper-and-pencil neighbors.[9] There is still skepticism about the spreading of cyberspace under the big sky. "The first four ranchers that I know of that started using computers," said one rancher, "all went broke within five years."[10] But when tax accountants, breeders' associations, suppliers, and county agents all computerize, the ranches cannot afford to remain islands of natural and cultural information. Some ranchers look forward to cyberherding. "When scanners can read top," said one, "and use a scale under a working chute then we will gather the info. Data gathering needs to be automated."[11] To the agricultural information industry this is a meek request. Information technology is eager to deliver "precision agriculture" where everything under the sun will be measured, monitored, and controlled.[12] As in business generally, whether productivity will justify the investment in computers is an open question though not one individual ranchers are at liberty to answer according to their lights. Information technology is sweeping everything before it. What in any event is likely to get lost is the symmetry of natural information and human competence that is reflected in this observation: "My husband knows his cattle personally by working with them and has a memory for traits, problems and style. His father had that trait and so he does and our son seems to have it also."[13]

A similar loss is taking place in the wilderness of Montana. Smoke Elser, Missoula's revered outfitter, knows the Bob Marshall Wilderness as well as anyone and can tell his clients any time just where they are on their trip. But he has been shown up, at least as far as accuracy is concerned, by a know-nothing dude who carried a global positioning system (GPS) receiver that told him within fifty or so feet where he was. Such a device can also track his progress, tell him how far he has traveled from the trailhead and how long it will take to the camp. And if he liked the trip, he can store all this information and retrace his trip exactly a few years hence, stopping at all of Smoke's favorite campsites and fishing spots.[14] Soon it will take deliberate recklessness to lose cattle without a trace or to get lost on a wilderness trip. The last dark and dangerous recesses of the world and the remaining burdens of how to become intimate with the land will appear to have become things of the past.

There are grizzlies and wolves in Yellowstone Park, but you may not want to encounter the former and are at any rate unlikely to see either species when you travel through the park. What an embarrassment, however, to return to New York having to confess to your friends that you saw not hide nor hair of the two most charismatic members of the megafauna. To prevent such calamity, the Grizzly Discovery Center has established itself at the west entrance to the park and exhibits grizzlies and wolves, contented and playful to all appearances, and yet, much like their human spectators, cut off from the environment that once engaged their skills and warranted their ferocious power. The IMAX theater next door will hourly show you *Yellowstone*, the movie, on a screen five stories high and half a block wide. Enveloped by symphonic music pouring forth from the fourteen speakers of a six-channel stereo surround system, you glide over the sunny expanses of the park, move through centuries of human history, penetrate the geology of the geysers, come face-to-face with eagles and bears. The real park must appear dreary and boring in comparison.

But why drive all the way from the east coast to begin with when the experience of the wild west is so much closer at Disney World's Wilderness Lodge? Inside it, Silver Springs Creek emerges from the floor, flows past wildflowers, tumbles down falls, empties into pools, continues on the far side of the grounds, and ends in Fire Rock Geyser, shooting up 180 feet every hour on the hour—more faithfully than Old Faithful ever does.[15] Thus Yellowstone Park has become negligible, and so have traditional city libraries, concert halls, and street musicians now that we can have a splendid cultural and urban experience in our home theater.

Technological information is the consummation of a development that began a century ago. Describing the three decades that followed the Civil War, Alan Trachtenberg noted that "[i]n technologies of communication, vicarious experiences began to erode direct physical experience of the world."[16] And still more to the point, he observed that "the more knowable the world came to seem as *information*, the more remote and opaque it came to be seen as *experience*."[17] In Bertrand Russell's terms, our knowledge by description has displaced our knowledge by acquaintance, and with the displacement of the latter the difference between the focal area of the nearness of things and the peripheral area of information about things has dissolved.[18] It has done so in cognition as well as in emotion and in labor as well as in leisure. In our serious dealings with the world, we generate and possess more information than ever. But reality itself gets ever more deeply buried under all the information we have about it. In consumption we are getting so adjusted to the light fare of more or less virtual experiences and emotions that the reality of persons and things seems offensively heavy and crude.

One might find this account of our situation too optimistic. Based on information technology, our omniscience and omnipotence have achieved such transparency and control of information that there are no things any more to be discovered beyond the signs. Nothing any longer is buried beneath information. Behind the virtual self-representations there are no real persons left to be acknowledged. The philosopher Martin Heidegger, who throughout his life had struggled to articulate a sense of being beyond human machination, has been scolded for his nostalgic "Yearning for Hardness and Heaviness."[19] And as if taking this lesson to heart, Bill McKibben has movingly lamented the passing of nature's unsurpassable sovereignty.[20]

Adjusting the Balance

There is no danger that technological information might entirely displace natural or cultural information. Inevitably we will spend much of our time navigating tangible environments, paying attention to the natural signs that tell us where we are going. Writing—the principal kind of cultural information—will remain indispensable as well. There is a real possibility, however, that natural and cultural information will decline to mere utilities, tools we need but fail to sustain as signs of irreplaceable kinds of excellence.

The succeeding kinds of information accomplish and in some ways surpass what their predecessors provide, and so it may seem as though technological information could give us the best of three worlds, the most powerful and refined information about, for, and as reality. But the successors do not fulfill their ancestors' task in quite the same way. Something does get lost when a later kind of information displaces an earlier. The loss is insignificant where natural or cultural information has irreversibly declined to a mere means or has reached the limits of its disclosive or transformative power. Most people, to be assured of a square meal, have no need of reading the tracks of animals or recognizing the edible fruits and plants in the wild. And no astronomer would want to be reduced to gathering information through optical telescopes and recording it by hand. It is otherwise when natural and cultural signs are at their highest.

For natural signs this is the case when they point us to a landmark and, having brought us face-to-face with it, mark off the presence and nearness of the focal area against the horizons of distance and the past. Nothing so engages the fullness of human capabilities as a coherent and focused world of natural information. No amount or sophistication of cultural or technological information can compensate for the loss of well-being we would suffer if we let the realm of natural information decay to one of resources, storage, and transportation. Analogously, nothing so concentrates human creativity and discipline as the austerity of cultural information, provided the latter again is of the highest order, consisting of the great literature of fiction, poetry, and music. Our power of realizing information and our competence in enriching the life of the mind and spirit would atrophy if we surrendered the task of realization to information technology. Perhaps what holds for the realization of information goes for its production too. The simplicity of the pen and the blankness of paper may be able to challenge, if they do not terrify, the resources of the writer and drafter in a way the obliging servility of the computer cannot match.[21]

Adjusting the kinds of information to one another and balancing information and reality cannot possibly be a return to earlier conditions. Modern technology and information technology particularly are tokens of a profound and irreversible change in the nature of reality. In the premodern world, the material force of reality issued in moral instruction and favored, though it did not guarantee, moral practices. Respectful attention to wild animals was a condition of survival for Native American hunters.[22] Marital fidelity, loyalty to one's band, and a tradition of courage allowed the Blackfeet to cope with harsh weather, scarce game, and injury or untimely death.[23] Within the postmodern lightness of being, to the contrary, the moral instruction of reality can restore its material force and with it our fundamental welfare—the full engagement of human capacities. We are essentially bodily creatures that have evolved over many hundreds of thousands of years to be mindful of the world not just through our intellect or our senses but through our very muscles and bones. We are stunting and ignoring this ancestral attunement to reality at our peril.[24]

It is impossible, however, to recover the full symmetry of humanity and reality through a virtual representation of reality. Information theory can be employed to show that, if virtual reality were to be made informationally exactly as rich as actual reality, the virtual would have to become a verbatim duplicate of the actual—a reduction to absurdity. But finally it is not the requirement of informational plenitude that leads us back to actuality but the moral eloquence of reality itself. As long as we remain in a cocoon of virtual reality or behold and control actual reality chiefly through information technology, the world out there seems light and immaterial. But once we take up the challenge of a natural area or the invitation of a truly urban space, material reality reappears in its commanding presence and engages our bodily exertion and spiritual pleasure to the limits of our capacities. But to repeat, the material vigor and splendor of reality will rest on its moral authority and our ability to respond to it. We can at any moment escape from the rigors of nature and the burdens of urbanity and surrender both city and country to neglect and abuse. The penalty of such evasiveness is no longer starvation or bodily injury but moral atrophy.

Righting the balance of information and reality is the crucial task. It amounts to the restoration of eminent natural information. A well-ordered realm of natural information in turn is both hospitable to practices of realizing cultural information and enlivened by such practices. As for technological information, there is no sense in trying to channel its development through narrow proscriptions or prescriptions. Nor does it make sense, of course, to let it run wild and overrun nature and culture. It is best allowed to develop freely within a world whose natural and cultural ecologies are guarded and engaged in their own right.

There are indications that people in this country are beginning to heed the voices of reality and not only are seeking engagement with things of nature and culture but are determined to renew what has been disfigured through neglect or brutality.

The Rattlesnake Valley in western Montana was settled by squatters and homesteaders only a hundred years ago. They found the valley in much the same condition it had been in for hundreds and perhaps a few thousand years. The settlers built log cabins and dug root cellars, cleared the valley bottom of trees and rocks, strung barbed wire fences, dug irrigation ditches, and did subsistence farming and ranching on thin and stingy soil. To make ends meet they logged the hillsides, pulling and shoving the logs down timber draws. The logs were used to make ties for the approaching Northern Pacific railroad and to supply the citizens of Missoula with firewood.[25]

Rattlesnake Creek was an early source of drinking water for Missoula. During the Great Depression the water company's desire to protect the watershed converged with the grinding poverty of the Rattlesnake farmers, and the company bought out the settlers in the upper region of the valley. The houses and barns were leveled, the timbers and implements hauled away; barbed wire and garbage is being removed to this day. Gradually, however, the water company shifted the source of its supply to the safer and more convenient wells in the Missoula Valley. Development of the easily accessible and spectacular building sites in the upper Rattlesnake would have been the normal course of events. But environmentalist Cass Chinske and Congressman Pat Williams rallied the sense of the citizenry that the area should be preserved for reasons the congressional act that secured the federal lands and provided for the acquisition of the private holdings put this way: "This national forest area has long been used as a wilderness by Montanans and by people throughout the Nation who value it as a source of solitude, wildlife, clean, free-flowing waters stored and used for municipal purposes for over a century, and primitive recreation, to include such activities as hiking, camping, backpacking, hunting, fishing, horse riding, and bicycling; and certain other lands on the Lolo National Forest, while not predominantly of wilderness quality, have high value for municipal watershed, recreation, wildlife habitat, and ecological and educational purposes."[26]

The preservation of a natural area can no longer be a matter of simple human withdrawal, nor can it seek guidance from a pretechnological norm of wilderness. It must take the form of a conversation with nature that seeks answers to questions like these: Should the noxious Eurasian weeds that are invading the valley be stopped by all available means? How far can wildfires be allowed to burn? Should eroding trails be reconstructed? Does it or did it make sense to reintroduce mountain goats, bighorn sheep, moose, fishers, wolves, or grizzly bears? Refusing to face these questions is merely one way of answering them.

Still, a natural area, however much it is informed by human decisions, leads a life of its own, and its ecology of things and economy of signs most of the time have a calmness and autonomy that set nature apart from culture. To enter a natural area is to be greeted and astounded by life in its own right. At the same time, the life of nature engages you most deeply if you understand it in the context of cultural information, that is, within the space of intelligibility that is circumscribed by information about the history of the place.

Thus walking north from the trailhead along Rattlesnake Creek, you come to the abutments of a long-gone bridge that Sebastian Effinger, the first settler in this area, threw across the river to get to the western part of his homestead, the meadows at the confluence of Spring Creek and the Rattlesnake. The road used to angle up from the bridge, and where it once reached the bench of the meadow you walk across flat stones, nearly invisible now, that used to support the schoolhouse. Trees are now invading the meadows that Effinger and his neighbors cleared. They felled the big ponderosas that had stood their ground when the periodic fires cleared out the understory. And they loaded the scattered stones on low sleds they called rock boats and dragged them off to the side where they now rise from the grass in low, grey piles. They denuded the hillsides as well, crisscrossed the valley bottom with dirt roads, and dumped their garbage and cans wherever they were out of the way. Early in this century, the valley must have looked ragged and disheveled.

Meanwhile, second growth ponderosa, Douglas fir, and western larch have come up among the few gigantic ponderosas that loggers have spared. Now the north facing slopes look lovely, dark, and deep. Those facing south are still open and sunny, but they too are "dougging in" as the foresters say. Walking on up along the Rattlesnake you find the remnants of a flume that used to carry water from a ditch across the creek to irrigate the meadows on the eastern side of the river. The ditches were dug by hand and where possible by mule-drawn scrapers, huge two-handled shovels that look like a wheelless wheelbarrow. The farmers used the same devices to raise the natural dams of the mountain lakes so that the water that accumulated in the spring could be released in the summer to replenish the Rattlesnake and the ditches it used to feed.

You undertake such backbreaking work in the hope of laying a foundation of prosperity for your children and grandchildren. The dams were helpful to Missoula's water supply for a couple of generations or so. But the ditches, roads, houses, root cellars, and barns turned out to be labor poorly spent. They have collapsed into scattered traces of desperately brave work, overgrown by the quiet of grass and shrubs, and shaded by apple trees gone wild on the abandoned homesteads. But mingled with admiration for the courage and with pity for the failures of one's forebears, there is gratitude to one's contemporaries for allowing nature to reclaim the once tortured land.

Moving further north, the old road rises up the Hogback, the terminal moraine of the Rattlesnake glacier that sits astride the valley and forces the river to the east slope where the Rattlesnake has cut an opening through the glacial wash. And so on up the valley through a series of portals that open on meadows until after some twenty miles you reach the high country of the wilderness area. When on a summer's morning you run up the valley, the sun rises over the east slopes like a blast of trumpets, the canyon walls open like the doors of a cathedral. Humans have acceded to nature, nature graces humans.

The rising sensitivity to the claims of nature is matched by a growing recognition of the splendor of the city and of the injuries cities have suffered.[27] The automobile has been the bane of traditional urbanity. Automobility, of course, is only a symptom of our commitment to technological liberty and is now so deeply embedded in the culture that there is no rolling it back. But we can hope to counter the distended monotony of the postwar suburbs and the forbidding brutality of city centers with urban spaces diverse enough to be engaging and articulate enough to be intelligible. It has been the concern of the New Urbanism to recreate the compactness and comprehensibility of traditional towns in new developments outside the city. The results surely have some of the coherence and rhythm of the natural ecology of things and signs—a scale that can be bodily appropriated, an ordering of the parts that indicates the sense of the whole, a placement of landmarks that provides orientation.[28]

The crucial challenge, however, is the one that the New Urbanism has turned to just now, the restoration of the central city itself.[29] There are enough remnants of earlier order and splendor to indicate what an urban space on a grand scale should be, a place anchored by commanding landmarks such as theaters, set in streets or squares that are convenient for pedestrians and inviting for people who want to sit, eat, read, or watch. Such spaces too disclose their meaning readily and variously, exhibiting signs that inconspicuously become things so that the balance of presence and reference remains intact. But more is needed than arranging for signs and things to emulate the sane and measured world of natural information. Such arrangements all by themselves can become thin and lifeless. Worrying about this kind of problem, Herbert Muschamp has warned that "[l]ike the modernists, the new urbanists rely too much on aesthetic solutions to the social problems created by urban sprawl."[30] Most of these problems are distressingly urgent and obvious—crime, poverty, unemployment, decay, and dirt. Pragmatism, however, is a better route toward their solution than grand theory. Though morally urgent and hard to reach, solutions to these problems would still leave us with the specter of urban spaces without depth and vitality, "architecture for the Prozac age," as Muschamp calls it, "Potemkin villages for dysfunctional families."[31]

Nor is it the case that the morally urgent and the culturally subtle problems can only be, or should be, solved sequentially—first safe streets then vital urbanism. In fact as matters stand, the latter promotes the former. There is a large measure of truth in Jane Jacobs's thesis that engaging and healthy city life is the ground state and that crime and decay move in only after true urbanity has moved out.[32] Moreover urban vitality not only enlivens aesthetically pleasing spaces; it can also mark out a place in otherwise inconspicuous space. Daily urban life at its best is regular in the large and improvised in the small. Its daily rhythm provides a framework for the small variations of shopping, strolling, conversing, and watching. The fabric of daily city life in turn provides the backdrop for those festive events that lend cultural life significant structure. On such an occasion space comes alive, time is focused, and people are inspired. The convergence and florescence of these elements we call celebration. In our culture, public and communal celebrations are for the most part realizations of cultural information. Thus the enlivening moments in contemporary life are fusions of natural and cultural information, the skilled performance of a score or play in a well-ordered and intelligible setting.

One such monumental moment took place on 23 June 1993, when Luciano Pavarotti sang on the Great Lawn of Central Park in New York City before a crowd of half a million. Naturally the park lends itself, and does so every day, to less focused and massive celebrations. But the presence of so many riveted on just one person draws the city together and quickens its life like few other events. This particular part of Manhattan's emerald jewel is more dusty than green. Yet bordered by trees and lawns and surrounded by a familiar skyline, it leaves you in no doubt where you are. Pavarotti's singing that evening made it plain why you would want to be there and why there should be a place like the Great Lawn. It was filled with the joys, sorrows, and the consoling pleasures of the Italian tenor repertoire. Though the scores were familiar, the contingency of the occasion was poignant. Pavarotti's career had been in disarray and his voice in decline. Who would prevail in the struggle between tenor and time? There might have been a shadow of doubt in the opening Verdi aria "Quando le sere al placido," but soon Pavarotti rose to the soaring brilliance of his younger years. When at length he concluded with Puccini's "Nessun Dorma," jubilation swept across the Great Lawn from the stage to the last ranks of listeners.

The occasion was ringed by the ambiguities of technological information as much as by groves and high rises. To most of the audience Pavarotti's voice came through loudspeakers the size of outhouses, hoisted up on fork lifts, and his presence was reduced to something the size of one's thumb. Ironically a big screen behind him displayed his face in hyperreal size and color. One could have had a far superior view and sound in front of one's television set. Or, if sound alone and supremely mattered, one might have waited for the hyperreal studio renditions of London Records, "drawn from its rich catalog of Pavarotti recordings."[33] Yet many thousands were drawn to the occasion where music became real then and there.

Three days later another concert took place in Manhattan at the inauspicious corner of 51st and Park. It was noon, and Randy and the Rainbows, the one-hit wonders of "Denise" fame, were doing golden oldies. Slowly people gathered, and soon there was a crowd of three hundred, overcome by the sweet sounds they once had poured their longings and passions into, "Only You," "Midnight Confession," "Don't Take Away the Music." Again they could have had better sound more readily from CDs and a stereo. But standing there, all more or less settled and accomplished, sadder than they once were and perhaps wiser, they could see this place and their lives within the space of their hopes and sorrows, and it was Randy's and the Rainbows' music that had summoned and enthralled them.

That evening, Lynn Redgrave once again presented *Shakespeare for My Father* in the Helen Hayes Theater on Broadway. Given the hypercharged entertainment we are used to getting from screens large and small, this seemed like an impossibly cool and austere affair, one actress telling stories and acting out scenes from Shakespeare. There was a particular grace to these hours, when a highly schooled and dedicated actress once again risked her stamina and reputation to gather for her audience a powerful strand of the theater's history. In her acting, pieces of literature, personages of the stage, and events of recent history were knitted together into a context that made sense of troubled circumstances, however implicitly and suggestively.

Much of Redgrave's material is available through information technology, and so are terabytes of closely connected information. Anyone familiar with the Web and the Net could have assembled an electronic portfolio that would be more varied and colorful than *Shakespeare for My Father*. But all this technological information is floating inconsequentially in cyberspace. It takes an actual time and place to gather an audience, it takes an audience to create a sense of expectation, it takes a real person to warrant such a concentration of reality and humanity, and finally it is such a warrant that sustains a place like the Helen Hayes Theater and made the architect Hugh Hardy restore the Victory Theater to its original splendor. Hardy is quite clear about the significance of his work and believes, as the *New York Times* had it,

> that it is the vitality of public spaces that keeps a city healthy enough to counter the mounting popularity of simulated experience—theme parks as well as enclosed shopping malls over the real thing.
>
> "People used to understand that gathering in public was good; that's what democracy meant," said Mr. Hardy.[34]

Good News

There are indications, then, that we are beginning again to recognize reality, to realize information, and to right the balance of signs and things. The symmetry of information and reality reaches its highest point in celebration when the ambiguities of signs and things complement and resolve each other. In celebration the austerity of a score or a text is redeemed through realization and the diffusion of reality is focused through a script. Information comes alive, reality becomes eloquent. When this happens, a celebration rises to the stature of a landmark in time, something that gives our lives coherence and significance. It marks a meaningful moment in the ravages of time. Like a landmark, such a moment of high contingency is both a sign and a thing. As a thing it has the presence of self-warranting clarity, as a sign it refers us to the darkness of contingency that constitutes its periphery.

A momentous celebration is suffused with enthusiasm and harmony. Half a million often surly New Yorkers gathered for pleasant and even artful picnics, made room for one another, exchanged amicable remarks, and were united in appreciation and affection when Pavarotti sang. But all this good will and pleasure were challenged, contradicted, and perhaps belied by the misery of the homeless nearby, by the muggings occurring in the run-down sections of Manhattan, by the strife, the diseases, and the starvation around the globe that every citizen of this country is somehow implicated in. The deplorable conditions that surround and question every celebration are no more distressing than the moral debilities that shadow individual joy. Surrounded by the peace and grandeur of the upper Rattlesnake Valley, I may feel the strength to forgive my enemies and to take on the labors of order and charity I have been avoiding. There is in fact no denying that resourcefulness and forbearance flow from natural and communal celebration. But at length I do fall prey again to anger and anxiety and unsay and undo the grace of nature or community. The warm and luminous moments of festivity are forever surrounded and threatened by darkness.

The darkness of contingency can open up in different ways. One is the violent ruination, the other the slow evaporation of meaning.

The former is a holocaust and leaves us with ashes, the latter is oblivion and leaves us with nothing. The holocaust is the catastrophe of real ambiguity when reality more than disperses its meaning and energetically ruins it, most terrifyingly through those vital concentrations of reality we call persons. Oblivion is the failure of symbolic ambiguity, the human inability to meet the rush of time with the stability of memorable signs.

The common response to devastation and oblivion is remembrance. We cannot redeem the holocaust, but we must remember it, and we try to do so by erecting enduring things and seeing to truthful signs. Remembrance is the response to the evanescence of meaning as well. To be forgotten or to see events disappear through forgetfulness is deeply, if much less painfully, disturbing to human beings. To escape the raging darkness of oblivion Alexander the Great took a flock of historians on his campaign but must have been uncertain of their skill, for when he came upon the grave of Achilles, he exclaimed, "Fortunate young man, to have found in Homer the herald of your valor."[35] Similarly, Ben Sira urged his community to embrace and remember its forebears:

> Let us now sing the praises of famous men,
> our ancestors in their generations.[36]

And he makes clear the consequences of the failure of remembrance:

> Some of them leave behind a name,
> so that others declare their praise.
> But of others there is no memory;
> they have perished as though they had never existed;
> they have become as though they had never been born,
> they and their children after them.[37]

Writing, of course, is the great invention that produces enduring records where memory would fail. The power of this instrument particularly impresses itself on cultural awareness when it is first introduced on a wide scale as it was in Norman England. Accordingly Henry III, concerned about the memory of the celebration of Edward the Confessor in 1247, addressed a chronicler from the throne and ordered him "to write a plain and full account of all these proceedings, and insert them in indelible characters in a book, that the recollection of them may not in any way be lost to posterity at any future ages."[38]

Information technology has deeply influenced the ways we cope today with the threat of the devastation and loss of meaning. The challenge to the festive resolution of the ambiguity that rises from the surrounding injustice and misery we are inclined to meet with a version of virtual ambiguity, a loosening of the ties that should connect our celebrations with their real and entire context. While virtuality is our reply to the devastation of common meanings, hyperinformation is our response to the oblivion of individuals. Common hyperinformation is the huge amount of colorful information we accumulate through pictures and videos especially. But all the other records we keep and that are kept about us are part of hyperinformation. Utopian hyperinformation is the brainchild of scientists who, in the tradition of artificial intelligence, believe that the core of an individual is the information contained in the brain, and purport that software can and will be extracted from the wetware of neurons and transferred without loss to the hardware of a computer or some other medium forever and again in this way and that so that the core of individuals, their personal identity, will achieve immortality.[39]

All of these are desperate attempts. To the extent that we shield celebration through virtual ambiguity from the reproaches of a suffering world, we empty celebration of meaning. At the limit, when celebration is fully protected, it is no longer worth saving. The endeavor to be remembered through common hyperinformation is indistinguishable from a headlong rush into oblivion. Some of the information will be overtaken by its physical and social fragility. If we find a way to stabilize it, however, its sheer disorganized and imposing mass will excuse our offspring from taking it to heart.

The reach for utopian hyperinformation is perhaps the most telling and melancholy indication that no one wants to "pass into oblivion," as a medieval chronicler has it, "as hail and snow melt in the waters of a swift river swept away by the current never to return."[40] In ordinary people, atheists or not, this fear takes the form of the desire to be remembered and to be remembered well. Alexander wanted more than to be recorded by historians, he wanted to be transfigured by the poet. Ben Sira asks us not just to recall but to praise our ancestors. People seem to conceive of themselves as deeply ambiguous signs that call for resolution.

What is true of the microcosm of the person is true of the macrocosm of the universe. It too is a sign as much as a thing, something that refers to its beginning and end. In our culture the normative response to the unresolved references of the cosmos is astrophysics. Though it is not inevitably committed to a beginning and an end of all things, it does concern itself with the lawful structure of the world's past and future and aims at a final theory of all there is.[41]

As regards the middle region of the cosmos that is so artfully balanced between the structure of atoms and galaxies, the terrestrial realm of nature and culture, its welfare requires more calmness and lucidity of recollection.[42] As it is, contemporary culture may lapse into a condition where a surfeit of information is as injurious as the lack of information. Where in the latter case one is confined by the darkness of ignorance and forgetfulness, today we are blinded by the glare of excessive and confused information. To regain our sight for the coherence of the public world we must be able to count on our chroniclers—the journalists, essayists, and historians—and we must allow their work to come to rest and attention for a day at least, or a month, or some years. Newspapers, journals, and books have been the places of considered judgment, and these or some such focal points are needed if information technology, beyond its instrumental functions in science and industry, is to become a constructive strand in the texture of our lives.

To recover a sense of continuity and depth in our personal world, we have to become again readers of texts and tellers of stories. Books have a permanence that inspires conversation and recollection. When you read or recount a passage from a book to your loved one, the matter at issue envelops both of you and fills the place you occupy.[43] Stories are the spaces wherein pictures and mementos come to life and coalesce into a coherent picture of the past and a hopeful vision of the future. Records in turn keep stories straight and lend them detail. Thus the culture of the word can card, spin, and knit the mass of technological information into a tapestry that is commensurate with reality.

As for cosmic closure, I quite agree with Steven Weinberg that a final theory will be a noble and intellectually satisfying accomplishment, the crowning achievement in our search for structure and lawfulness.[44] But the world has a history as well as structure. History in the large and strict sense is the meaningful sequence of unpredictable events. It is contingency. Hence we face the question whether there can be cosmic closure of a historical as well as a structural sort. It may well turn out that at the beginning of the cosmos history and structure are indistinguishable, that the unfolding of the cosmos is simultaneously an unfolding of lawfulness. But in time structure and contingency must diverge in one way or another. There is no prospect of deducing the lightning that causes a devastating fire or the encounter that leads to a happy marriage from the laws of the universe alone.

History, then, requires its own kind of reading, one that must consist with the laws of nature but also attend to the givenness of things and events. The decline of meaning and the rise of information have kept contemporary readings of history weak and inconclusive. The recent burst of information technology has further, and fortunately, silenced the voices of overt misery, of disease, poverty, and violence, both here and around the globe. There is still unspeakable suffering in many parts of the world. But information technology is both the channel and the energy that is carrying the free market economy and its blessings to every corner on earth.

As overt misery is waning, so is the inference that used to be drawn from it, namely, that suffering would not be what it is if it did not intimate salvation in the end. And similarly, as our celebrations are losing their context and contrast of poverty and violence, they also lose their reference, weak already, to the need for final salvation. But while information technology is alleviating overt misery, it is aggravating a hidden sort of suffering that follows from the slow obliteration of human substance. It is the misery of persons who lose their well-being not to violence or oblivion, but to the dilation and attenuation they suffer when the moral gravity and material density of things is overlaid by the lightness of information. People are losing their character and definition in the levity of cyberspace.

The engagement of reality is the proximate remedy for this condition, and yet many of us find it hard to face up and to be faithful to persons and things. Though we feel blessed by celebrations once we have been drawn into them, all too often we lack the strength or loyalty to enter them regularly. The moral paralysis people inflict on themselves through the abuse of technological information is miserable any way you look at it. The constructive responses are manifold, however, and not a matter of contestation but attestation. Christians, for example, owe what fidelity to persons and festive things they possess to a strong reading of cosmic contingency—the history of salvation. Whatever definition they attain as persons through their engagement with reality they see as precarious and in need of final resolution. The world as a sign makes them look forward to the event when

> Liber scriptus proferetur,
> In quo totum continetur,
> Unde mundus judicetur.[45]

> A written book will be brought forward
> Wherein everything is gathered
> Whence the world may be adjudged.

All of us will be remembered and more; our souls will be rocked in the bosom of Abraham.

Introduction: Information vs. Reality

1. Brent Staples, "Life in the Information Age," *New York Times*, 7 July 1992.

2. Michael L. Dertouzos, "Communications, Computers, and Networks," *Scientific American*, September 1991, 62.

3. Bill Powell, "Eyes on the Future," *Newsweek*, 31 May 1993, 39.

4. Rudy Rucker, *Exploring Cellular Automata* (Sausalito, Cal.: Autodesk, 1989), 16.

5. Jay David Bolter, *Writing Space: The Computer, Hypertext, and the History of Writing* (Hillsdale, N.J.: Erlbaum, 1991); Richard A. Lanham, *The Electronic Word: Democracy, Technology, and the Arts* (Chicago: University of Chicago Press, 1993).

6. Johann Wolfgang von Goethe, "Der Zauberlehrling," in *Goethes Werke* (Hamburg: Wegner, 1948), 1: 276–79.

7. Bernard Sharratt, "Please Touch the Paintings," *New York Times Book Review*, 6 March 1994, 3.

8. Rev. 21:3, 7–8; 21:1. These and the following citations from the Bible are from the King James translation.

9. "The Second Coming," in *The Collected Poems of W. B. Yeats* (New York: Macmillan, 1959), 184–85.

Chapter One: The Decline of Meaning and the Rise of Information

1. Claude E. Shannon, "The Mathematical Theory of Communication [1948]," in Shannon and Warren Weaver, *The Mathematical Theory of Communication* (Urbana: University of Illinois Press, 1949), 3–91.

2. Fritz Machlup, *The Production and Distribution of Knowledge in the United States* (Princeton: Princeton University Press, 1962), 8. Eventually Machlup gave information his full attention and commissioned accounts of its various fields to aid him in the recasting of his work on knowledge. These accounts, along with a prologue and an epilogue by Machlup, were published as Machlup and Una Mansfield, eds., *The Study of Information: Interdisciplinary Messages* (New York: Wiley, 1983).

3. Matthys Levy and Mario Salvadori, *Why Buildings Fall Down: How Structures Fail* (New York: Norton, 1992), 209.

4. Bryan Appleyard, *Understanding the Present: Science and the Soul of Modern Man* (New York: Doubleday, 1993).

5. One focus of this debate is supervenience. See, e.g., Terence Horgan, "From Supervenience to Superdupervenience: Meeting the Demands of the Material World," *Mind* 102 (1993): 555–86.

6. Robert Wright, *Three Scientists and Their Gods: Looking for Meaning in an Age of Information* (New York: Harper, 1989), 4; Keith Devlin, *Logic and Information* (Cambridge: Cambridge University Press, 1991), 2.

7. Donald M. MacKay, *Information, Mechanism, and Meaning* (Cambridge: MIT Press, 1969), 160.

8. To the contrary see Dean W. Zimmerman, "Could Extended Objects Be Made Out of Simple Parts? An Argument for 'Atomless Gunk,'" *Philosophy and Phenomenological Research* 51 (1996): 1–29.

9. Wright, *Three Scientists*, 3–80.

10. It is evidently helpful to the discussion of black holes and in particular of the question whether or not (structural) information gets irretrievably lost in black holes. See Stephen W. Hawking and Roger Penrose, *The Nature of Space and Time* (Princeton: Princeton University Press, 1996).

11. Wright, *Three Scientists*, 90.

12. Fertility and rigor are two different things, however. See Jeff Coulter, "The Informed Neuron: Issues in the Use of Information Theory in the Behavioral Sciences," *Minds and Machines* 5 (1995): 583–96.

13. William H. Ittelson, *The Ames Demonstrations in Perception* (with an interpretative manual by Adelbert Ames Jr.) (New York: Hafner, 1968), 170.

14. Ittelson, *Ames Demonstrations*, 41–42, 49–53, 183–96.

15. Anthony Bannon, "The Impact of Ames on Twentieth Century Art Theory and Criticism," *Proceedings of the Dartmouth Eye Institute Commemorative Symposium, 1994* (Hanover, N.H.: Dartmouth College), 163–74.

16. Weaver, "Recent Contributions to the Mathematical Theory of Communication," in Shannon and Weaver, *Mathematical Theory*, 98.

17. Bertrand Russell, "Knowledge by Acquaintance and Knowledge by Description," *Proceedings of the Aristotelian Society*, New Series 11 (1910–11), 108.

18. Russell, it must be noted, had an unduly narrow conception of acquaintance and presentation—sense-data can be known by acquaintance, but physical objects and other minds cannot.

19. J. L. Austin, *Sense and Sensibilia* (New York: Oxford University Press, 1962), 115.

20. David Israel and John Perry, "What is Information?" *Information, Language, and Cognition*, ed. Philip H. Hanson (Vancouver: University of British Columbia Press, 1990), 4.

21. William J. Mitchell, *City of Bits: Space, Place, and the Infobahn* (Cambridge: MIT Press, 1995), 24.

Chapter Two: The Nature of Information

1. For my understanding of information I am heavily indebted to Fred I. Dretske, *Knowledge and the Flow of Information* (Cambridge: MIT Press, 1981) and to Keith Devlin (and indirectly to Jon Barwise), *Logic and Information* (Cambridge: Cambridge Uni-

versity Press, 1991). However, my approach is broader and less formal than Dretske's. For my reservations about Devlin, see the following note. More immediately, I am grateful to Gordon G. Brittan Jr., Hubert Dreyfus, and Gene Moriarty for counsel and encouragement.

2. What is the structure of the message? And what is the nature of the ties between PERSON, SIGN, and THING? Devlin has attempted a rigorous answer to the former question under the heading of the "infon" and to the latter under the heading of "constraints." Devlin's valiant work has persuaded me, however, that information is so complex that a formal analysis that goes beyond some rough indications and aims at thoroughness (not to mention exhaustiveness) begins to resemble an increasingly vacuous redescription of the untidy distinctions we make in ordinary language or the unilluminating distinctions we make in physics and chemistry. For a critique of the school of thought Devlin belongs to, see Terry Winograd, "Moving the Semantic Fulcrum," *Linguistics and Philosophy* 8 (1985): 231–64.

3. This is adapted from Douglas R. Hofstadter, *Gödel, Escher, Bach* (New York: Basic Books, 1979), 335. Playfulness is made to perform useful work in "Multifunctioning Graphical Elements" in Edward R. Tufte, *The Visual Display of Quantitative Information* (Cheshire, Conn.: Graphics Press, 1983), 139–59. In the terminology of Nelson Goodman, *Languages of Art: An Approach to a Theory of Symbols* (Indianapolis: Hackett, 1976), 57–67, the oscillation between reference and presence is one between denotation and exemplification.

4. As Hollander remarks in *Types of Shape* (New Haven, Conn.: Yale University Press), x–xi, Lewis Carroll has a shaped poem with a pun on tale and tail in *Alice's Adventures in Wonderland, The Complete Works of Lewis Carroll* (London: Nonesuch, 1939), 35.

5. Don Schwenessen, "To Start a Fire," *Missoulian*, 9 October 1985; "Cash-cum-Kindling Keeps Skiers Warm," *Missoulian*, 30 January 1990.

6. John Chadwick, *The Decipherment of Linear B* (Cambridge: Cambridge University Press, 1958). Linear A, mainly due to the dearth of material, is undeciphered to this day. See David Schneider, "Pot Luck: Linear A, an Ancient Script, is Unearthed in Turkey," *Scientific American*, July 1996, 20.

7. Aristotle, *On the Soul* 3.8.

8. Gilbert Ryle sketched this problem in *The Concept of Mind* (New York: Barnes and Noble, 1949) where he talks about "The Systematic Elusiveness of the 'I,'" pp. 195–98. The conundrum was thoroughly investigated by Karl Popper in "Indeterminism in Quantum Physics and in Classical Physics," *British Journal for the Philosophy of Science* 1 (1950): 117–33, 173–95.

9. For accounts of such zooming in (or out), see Kees Boeke, *Cosmic View: The Universe in Forty Jumps* (New York: John Day, 1957) and Philip Morrison and Phylis Morrison, *Powers of Ten: A Book about the Relative Size of Things in the Universe and the Effect of Adding Another Zero* (New York: Scientific American Library, 1982).

10. Unlike John Haugeland, I can see no "mystery of original meaning." See his *Artificial Intelligence: The Very Idea* (Cambridge: MIT Press, 1985), 26–27, 89.

11. Donald Davidson, "The Folly of Trying to Define Truth," *Journal of Philosophy* 93 (1996): 263–78.

Chapter Three: Ancestral Information

1. Stephen R. Kellert, "The Biological Basis for Human Values of Nature," in *The Biophilia Hypothesis*, ed. Kellert and E. O. Wilson (Washington, D.C.: Island Press, 1993), 45–52.

2. Stephen Kaplan and Janet Frey Talbot, "Psychological Benefits of Wilderness Experience," in *Behavior and the Natural Environment*, ed. Irwin Altman and Joachim F. Wohlwill (New York: Plenum Press, 1983), 189–90.

3. Ella E. Clark, *Indian Legends from the Northern Rockies* (Norman: University of Oklahoma Press, 1966), 84–86.

4. For the fate of the trail to the buffalo, see Arthur L. Stone, *Following Old Trails* (1913; reprint, Missoula, Mont.: Pictorial Histories, 1996), 97–102; and Dan Oko, "State Neglects Ancient Indian Trail in Salvage Timber Sale on Blackfoot," *Missoula Independent*, 22–29 August 1996.

5. Harriet Miller and Elizabeth Harrison, *Coyote Tales of the Montana Salish* (Browning, Mont.: Museum of the Plains Indian, 1974), 12.

6. Rachel Kaplan, "The Role of Nature in the Urban Context," in *Behavior*, ed. Altman and Wohlwill, 154–55.

7. Clark, *Indian Legends*, 121–27.

8. Bill McKibben, *The Age of Missing Information* (New York: Random House, 1992); Kellert, "The Biological Basis," 49–53; and Roger S. Ulrich, "Aesthetic and Affective Responses to Natural Environment," in *Behavior*, ed. Altman and Wohlwill, 119–20.

9. McKibben, *Age of Missing Information*, 126.

10. Ibid., 129–30.

11. Jack Turner, "The Abstract Wild," *Witness* 3 (1989): 88; Kaplan and Talbot, "Psychological Benefits," 193–201.

12. Carl G. Hempel, "Aspects of Scientific Explanation," in *Aspects of Scientific Explanation* (New York: Free Press, 1965), 334.

13. Diana Raffman, "Toward a Cognitive Theory of Musical Ineffability," *Review of Metaphysics* 41 (1988): 693.

14. Fred I. Dretske, *Knowledge and the Flow of Information* (Cambridge: MIT Press, 1981), 137.

15. Keith Devlin, *Logic and Information* (Cambridge: Cambridge University Press, 1991), 16.

16. Dretske, *Knowledge*, 141.

17. Devlin, *Logic*, 150. See also pp. 20, 27, 151, and Dretske, *Knowledge*, p. 61. Further examples appear in A. C. Gattrell, "Concepts of Space and Geographical Data," in *Geographical Information Systems*, ed. David Maguire, Michael F. Goodchild, and David W. Rhind (New York: Longman, 1991), 124; Steven Pinker, *The Language Instinct* (New York: Morrow, 1994), 154 (Pinker, remarkably, is a realist); and John Horgan, *The End of Science: Facing the Limits of Knowledge in the Twilight of the Scientific Age* (New York: Addison Wesley, 1996), 228.

18. Devlin, *Logic*, 151.

19. J. Baird Callicott, "Traditional American Indian and Western European Attitudes toward Nature: An Overview," *Environmental Ethics* 4 (1982): 317.

20. Richard White, "Animals and Enterprise," in *The Oxford History of the American*

West, ed. Clyde A. Milner II, Carol A. O'Connor, and Martha A. Sandweiss (New York: Oxford University Press, 1994), 236.

21. Clark, *Indian Legends*, 82–83; Stone, *Following Old Trails*, 179–86. See also Callicott, "Traditional American Indian and Western European Attitudes," 301.

22. Clark, *Indian Legends*, 66–70; Miller and Harrison, *Coyote Tales*, 12–14.

23. A multitude of causal lines connects thing, sign, and recipient. Scientifically explicated causality, however, underdetermines meaning and information. Hence we must acknowledge terms like *eloquence, reality, intelligence, presence, reference,* etc. as primitive.

24. Clark, *Indian Legends*, 78.

25. Ibid., 77.

26. Paul Grice, "Meaning," in *Studies in the Way of Words* (Cambridge: Harvard University Press, 1989), 213–23.

27. Milo McLeod and Douglas Melton, *The Prehistory of the Lolo and Bitterroot National Forests* (Missoula, Mont.: Lolo National Forest, 1986), 6: 10–11.

28. Carling Malouf, "Stone Piles," *Archaeology in Montana* 3 (1962): 1–5; and "The 'Indian Post Office,'" *Archaeology in Montana* 5 (1964): 13–14.

29. John Van Seters, *Abraham in History and Tradition* (New Haven, Conn.: Yale University Press, 1975), 5–122.

30. Gen. 12:1–4.

31. Gen. 18:1–15.

32. Gen. 15:17–21.

33. Exod. 33:20–23.

34. 1 Kings 19:11–13. See also Granville C. Henry, *Forms of Concrescence: Alfred North Whitehead's Philosophy and Computer Programming Structures* (Lewisberg, Penn.: Bucknell University Press, 1993), 125–26.

35. Steven Weinberg, *Dreams of a Final Theory* (New York: Pantheon, 1992), 256.

36. Daniel C. Dennett, *Darwin's Dangerous Idea: Evolution and the Meanings of Life* (New York: Simon and Schuster, 1995), 154.

37. Weinberg, *Dreams*, 250.

38. Dennett, *Darwin's Dangerous Idea*, 520.

39. Victor Hugo, *Notre-Dame de Paris,* tr. Isabel F. Hapgood (New York: Crowell, 1888), 1:192.

40. Gen. 12:7.

41. Gen. 12:8; 13:18.

42. Gen. 26: 24–25; 35:1–7.

43. See notes 27 and 28 above.

44. Liz Bryan, *The Buffalo People: Prehistoric Archaeology on the Canadian Plains* (Edmonton: University of Alberta Press, 1991), 55–64; Andrew Nikiforuk, "Sacred Circles," *Canadian Geographic,* July/August 1992, 50–60.

45. Gen. 13:14–18.

46. James Welch, *Fools Crow* (New York: Viking, 1986), 3.

47. James D. Keyser, *Indian Rock Art of the Columbia Plateau* (Seattle: University of Washington Press, 1992).

48. For a brief summary of the scholarship on J and a spirited and improbable contribution to that scholarship along with a translation by David Rosenberg, see Harold Bloom, *The Book of J* (New York: Grove Weidenfeld, 1990).

Chapter Four: From Landmarks to Letters

1. Walter J. Ong, *Orality and Literacy: The Technologizing of the Word* (London: Methuen, 1982), 5–77.

2. Gen. 23:17–18. See also Gerhard von Rad, *Genesis: A Commentary*, tr. John H. Marks (Philadelphia: Westminster, 1961), 240–44; and John Van Seters, *Abraham in History and Tradition* (New Haven, Conn.: Yale University Press), 98–100.

3. M. T. Clanchy, *From Memory to Written Record: England 1066–1307* (Cambridge: Harvard University Press, 1979), 56, 232–33.

4. Ong, *Orality*, 67–68.

5. *The Making of Homeric Verse: The Collected Papers of Milman Parry*, ed. Adam Parry (Oxford: Clarendon Press, 1971); Rosalind Thomas, *Oral Tradition and Written Record in Classical Athens* (Cambridge: Cambridge University Press, 1989), 15–34.

6. Thomas, *Oral Tradition*, 55–59.

7. Clanchy, *From Memory*, 203.

8. Ibid., 208.

9. Ibid., 21.

10. Ibid., 24.

11. John D. Barrow, *Pi in the Sky: Counting, Thinking, and Being* (Boston: Little, Brown, 1992), 28–33; Denise Schmandt-Besserat, *Before Writing* (Austin: University of Texas Press, 1992), 1:158–61.

12. David Diringer, *The Alphabet: A Key to the History of Mankind*, 3d ed. (London: Hutchinson, 1968), 1:7–8, 2:12; Clanchy, *From Memory*, 88; Schmandt-Besserat, *Before Writing*, 1:161.

13. Clanchy, *From Memory*, 32, 95, pl. 8.

14. Schmandt-Besserat, *Before Writing*, 1:160, 185.

15. Thorkild Jacobsen, "The Relative Role of Technology and Literacy in the Development of Old World Civilizations," in *Human Origins*, 2d ed. (Chicago: University of Chicago Press, 1946), 145; Diringer, *The Alphabet*, 1:5, 2:1.

16. A. Seidenberg, "The Ritual Origin of Counting," *Archive for History of Exact Sciences* 2 (1962): 11–13.

17. Gen. 1:27.

18. Gen. 7:9.

19. Barrow, *Pi in the Sky*, 55.

20. Seidenberg, "The Ritual Origin of Counting," 8–9, 13–14.

21. Ibid., 3; Barrow, *Pi in the Sky*, 51–56; Schmandt-Besserat, *Before Writing*, 1:184–85.

22. Gen. 31: 44–46.

23. Ella E. Clark, *Indian Legends from the Northern Rockies* (Norman: University of Oklahoma Press, 1966), 110.

24. Ong, *Orality*, 85–93; Schmandt-Besserat, *Before Writing*, 1:1.

25. Ibid., 1:196.

26. Ibid., 1:6.

27. Ibid., 1:164.

28. Ibid., 1:197.

29. Jacob Klein, *Greek Mathematical Thought and the Origin of Algebra*, tr. Eva Brann (Cambridge: MIT Press, 1968), 46–48; Barrow, *Pi in the Sky*, 38–39; Schmandt-Besserat, *Before Writing*, 1:185–87.

30. Ibid., 1:8–10, 164–65, 191–92.

31. Ibid., 1:192.

32. Ibid., 1:194.

33. Ibid., 1:192.

34. I. J. Gelb, "Principles of Writing Systems within the Frame of Visual Communication," in *Processing of Visible Language 2*, ed. Paul A. Kolers, Merold E. Wrolstad, and Herman Bouma (New York: Plenum, 1980), 13–14; Diringer, *The Alphabet*, 1:17–112.

35. Chinese, in fact, has 50,000 to 80,000 characters. But many of them are really compounds. There is a stock of 200 to 500 components. See Diringer, *The Alphabet*, 1:13, and I. J. Gelb, *A Study of Writing* (Chicago: University of Chicago Press, 1963), 115–18. Chinese, moreover, is in part phonetic. See John DeFrancis, *The Chinese Language: Fact and Fantasy* (Honolulu: University of Hawaii Press, 1984), 133–48. Finally, the labor of learning logographic writing is balanced in part by the ease of logographic reading. Readers of alphabetic writing spontaneously take written words to be logographs and need to be taught laboriously how to analyze a word into its (more or less phonetic) letters. See Paul Bertelson, "The Onset of Literacy: Liminal Remarks," in *The Onset of Literacy: Cognitive Processes in Reading Acquisition*, ed. Bertelson (Cambridge: MIT Press, 1987), 8–9, 12.

36. Roy Harris, *The Origin of Writing* (LaSalle, Ill.: Open Court, 1986), 149.

37. Jacobsen, "The Relative Roles," 145; Schmandt-Besserat, *Before Writing*, 1:194.

38. Diringer, *The Alphabet*, 1:161–62.

39. Ibid., 1:163, 169.

40. N. H. Tur-Sinai, *Who Created the Alphabet?* (1949; reprint, Tel Aviv: Pales, n.d.).

41. It is controversial whether the North Semitic script is an early and vowelless alphabet (Diringer, *The Alphabet*, 1:158–70) or a late and truncated syllabary (Eric A. Havelock, *The Literate Revolution in Greece and Its Cultural Consequences* [Princeton: Princeton University Press, 1982], 66–88).

42. Diringer, *The Alphabet*, 1:358–64, 386–90, 418–32.

Chapter Five: The Rise of Literacy

1. Eric A. Havelock, *Preface to Plato* (Cambridge: Harvard University Press, 1963), 38–41; *The Muse Learns to Write: Reflections on Orality and Literacy from Antiquity to the Present* (New Haven, Conn.: Yale University Press, 1986), 87.

2. Denise Schmandt-Besserat, *Before Writing* (Austin: University of Texas Press, 1992), 1:1–4.

3. Plato, *Phaedrus* 275A–B. See also Plato, *Seventh Letter* 341E. Some of the translations are mine. For the most part they are from the Loeb Classical Library. As it turned out, written records and writing materials remained too scarce to disburden memory and obviate a good memory. In fact, more than a hundred years before Plato, Simonides of Ceos (556–486 B.C.E.) had established an art of memory that continued to flourish up until the eighteenth century. See Frances A. Yates, *The Art of Memory* (Chicago: University of Chicago Press, 1966). Printing and paper at length furnished the abundance of records and the opportunities for recording that made a well-trained memory unnecessary (Yates, xi, 4). Yet the question remains whether it was the laboriously acquired "artificial memory" that concerned Plato. More likely Plato scorned it since it was a concern of the Sophists. What worried Plato was the enfeeblement of memory as the enduring bond to reality, to the reality of the ideas most especially. See Yates, 37–39.

4. Havelock, *Preface to Plato*, 40; J. D. Denniston, "Technical Terms in Aristophanes," *Classical Quarterly*, July–October 1927, 117–18.

5. Plato, *Phaedrus* 276B–D. See also Plato, *Phaedrus* 277E and *Seventh Letter* 341C, 344C–D.

6. Plato, *Seventh Letter* 344C–D; Plato, *Phaedrus*, 277D, 278C–D.

7. Plato, *Laws* 769A; Plato, *Statesman* 294A.

8. Plato, *Seventh Letter* 341E.

9. Rosalind Thomas, *Oral Tradition and Written Record in Classical Athens* (Cambridge: Cambridge University Press, 1989), 60–61.

10. Thomas, *Oral Tradition*, 40.

11. Ibid., 38–40, 60–83.

12. Folkmar Thiele, *Die Freiburger Stadtschreiber im Mittelalter* (Freiburg: Wagnersche Universitätsbuchhandlung, 1973), 17–20.

13. Ibid., 69–72, 75–76.

14. Ibid., 25, 75–76. See also M. T. Clanchy, *From Memory to Written Record* (Cambridge: Harvard University Press, 1979), 139–41.

15. Thomas, *Oral Tradition*, 15–94.

16. Clanchy, *From Memory*, 17, 121.

17. Ibid., 7, 18–20.

18. Ibid., 1, 34–35.

19. Ibid., 37.

20. Ibid., 27, 123.

21. Ibid., 209.

22. 2 Cor. 3:6; Clanchy, *From Memory*, 210, 233–34.

23. Ibid., 233.

24. James Welch, *Fools Crow* (New York: Viking, 1986), 284.

25. Ibid., 383–84.

26. A. B. Guthrie, *Fair Land, Fair Land* (Boston: Houghton, 1982), 261.

27. Albert J. Partoll, "The Flathead Indian Treaty of 1855," *Pacific Northwest Quarterly*, July 1938, 292–93. For another account of the negotiations, see Arthur L. Stone, *Following Old Trails* (1913; reprint, Missoula, Mont.: Pictorial Histories, 1996), 76–82.

28. Plato, *Seventh Letter* 341C–D.

29. Elizabeth Ware Pearson, ed., *Letters from Port Royal, 1862–1868* (New York: Arno, 1969), ii, 67.

30. Plato, *Phaedrus* 275C. See also 276D.

31. Alfred North Whitehead, *Process and Reality* (New York: Harper, 1960), 63.

32. Friedrich Nietzsche, *Werke*, ed. Karl Schlechta (Munich: Hanser, 1966), 1:918 (my translation). Up until the Middle Ages, authors frequently dictated rather than wrote. The sculptural qualities of writing come fully into their own when authors are also writers. In antiquity, even authors who wrote dictated to themselves in a low voice. See Paul Saenger, "Silent Reading: Its Impact on Late Medieval Script and Society," *Viator* 13 (1982): 371–72.

Chapter Six: Producing Information:
Writing and Structure

1. Cultural information, in the sense I use the term, flourished in antiquity as well as in the Middle Ages and the early modern period, and in Asia as well as in Europe.

Setting aside limits of competence, any of these eras and areas could serve, if with less continuity and economy, to elucidate cultural information.

2. *The Sketchbook of Villard de Honnecourt*, ed. Theodore Bowie (Westport, Conn.: Greenwood, 1982), 106; *Villard de Honnecourt: Kritische Gesamtausgabe des Bauhütten-buches ms. 19093 der Pariser Nationalbibliothek*, 2d ed. (Graz, Austria: Akademische Druck- und Verlagsanstalt, 1972), fig. 29, 69–73.

3. Eric A. Havelock, *Preface to Plato* (Cambridge: Harvard University Press, 1963).

4. See particularly Plato, *Republic*, book 10 (595A–620D).

5. Plato, *Philebus* 18C.

6. The type-token distinction goes back to Charles Sanders Peirce, *Collected Papers*, ed. Charles Hartshorne and Paul Weiss (Cambridge: Harvard University Press, 1933), 4:423–24.

7. Plato, *Philebus* 18C.

8. Plato, *Theaetetus* 202B; *Cratylus* 423E, 425D.

9. Plato, *Theaetetus* 201E–202C.

10. An analogous problem arose in the counting system of tokens where, as Denise Schmandt-Besserat notes, "tokens to count sheep were supplemented by special tokens to count rams, ewes, and lambs. This proliferation of signs was bound to lead the system to its downfall." *Before Writing* (Austin: University of Texas Press, 1992), 1:162.

11. Plato, *Timaeus* 53A–56B. And what of the dodecahedron? "God used it up for the Universe in his decoration thereof." See *Timaeus* 55C.

12. To the extent that language is a "discrete combinatorial system," it works top down, rather than bottom up—a sentence is progressively articulated into phrases, words, symbols, and sounds. In any event, the point is lawful structure. See Steven Pinker, *The Language Instinct* (New York: Morrow, 1994), 83–191.

13. For an overview, see Pinker, *Language Instinct*, 83–125.

14. Aristotle, *On Interpretation* 16a1–9.

15. Plato, *Cratylus* 425A–B.

16. Plato, *Cratylus* 426C–427D.

17. Schmandt-Besserat, *Before Writing*, 1:161.

18. Ragnar Fjelland, "A Constructivist Foundation of Geometry," in *Praxeology*, ed. Fjelland and Gunnar Skirbekk (Oslo: Universitetsforlaget, 1983), 165–71.

19. Schmandt-Besserat, *Before Writing*, 1:162.

20. Fjelland, "Constructivist Foundation," 168; Seton Lloyd, "Building in Brick and Stone," in *A History of Technology*, ed. Charles Singer, E. J. Holmyard, and A. R. Hall (Oxford: Clarendon, 1954), 1:456–75.

21. Frits Staal, *Agni: The Vedic Ritual of the Fire Altar*, 2 vols. (Berkeley, Cal.: Asian Humanities Press, 1983).

22. A. Seidenberg, "The Origin of Mathematics," *Archive for History of Exact Sciences* 18 (1978): 331.

23. Seidenberg, "Origin of Mathematics," 325–26, 333–34; "The Ritual Origin of Geometry," *Archive for History of Exact Sciences* 1 (1960–62): 515–17; "Geometry of the Vedic Rituals," in Staal, *Agni*, 2:110–11.

24. What follows is a simplified version of Seidenberg's accounts in "Ritual Origin of Geometry," 490–91; "Origin of Mathematics," 320–21; and "Geometry of the Vedic Rituals," 96.

25. Seidenberg, "Ritual Origin of Geometry," 492–95.

26. Seidenberg, "Origin of Mathematics," 325, 327–28.

27. *The Thirteen Books of Euclid*, tr. and ed. T. L. Heath (Cambridge: University Press, 1908), 1:259, 343.

28. *Books of Euclid*, 1:129.

29. In book 1, proposition 47, in *Books of Euclid*, 1:349–50.

30. *Books of Euclid*, 1: 349.

31. Gordon G. Brittan Jr., "Constructibility and the World-Picture," in *Proceedings: Sixth International Kant Congress*, ed. G. Funke and Th. M. Seebohm (State College: Pennsylvania State University Press, 1985), 65.

32. Brittan, "Constructibility," 65–82, and "Algebra, Constructibility, and the Indeterminate," in *Causality, Method, and Modality*, ed. Brittan (Boston: Kluwer, 1991), 99–123.

33. Seidenberg, "Origin of Mathematics," 301–3.

34. Morris Kline, *Mathematical Thought from Ancient to Modern Times* (New York: Oxford University Press, 1972), 302–24.

35. Jacob Klein, *Greek Mathematical Thought and the Origin of Algebra*, tr. Eva Brann (Cambridge: MIT Press, 1994), 68.

36. Richard Hadden, *On the Shoulders of Merchants: Exchange and the Mathematical Conception of Nature in Early Modern Europe* (Albany: State University of New York Press, 1994), 68.

37. Klein, *Greek Mathematical Thought*, 204.

Chapter Seven: Producing Information:
Measures and Grids

1. Ludwig Wittgenstein, *Tractatus Logico-Philosophicus*, with English trans. by D. F. Pears and B. F. McGuinness (London: Routledge, 1961).

2. This program had a renaissance in the 1950s and 1960s under the heading of ideal-language philosophy. For discussion see Richard Rorty, ed., *The Linguistic Turn: Recent Essays in Philosophical Method* (Chicago: University of Chicago Press, 1967), 125–71.

3. Steven Pinker, *The Language Instinct* (New York: Morrow, 1994).

4. Keith Devlin, *Logic and Information* (Cambridge: Cambridge University Press, 1991), 28.

5. Diana Raffman, "Toward a Cognitive Theory of Musical Ineffability," *Review of Metaphysics* 41 (1988): 694.

6. Lloyd A. Brown, *The Story of Maps* (Boston: Little Brown, 1949), 35.

7. Brown, *Story of Maps*, 58–80.

8. Darrell Haug Davis, *The Earth and Man* (New York: Macmillan, 1950), 625.

9. The *boustrophedon* so called. See David Diringer, *Writing* (New York: Praeger, 1962), 56–58.

10. More precisely, the principal meridian is at 111 degrees, 39 minutes, and 33 seconds west; the base line is at 45 degrees, 47 minutes, and 13 seconds north. See *Manual of Instructions for the Survey of Public Lands in the United States* (Washington, DC: United States Department of the Interior, 1973), 60.

11. Brown, *Story of Maps*, 297; Dava Sobel, *Longitude: The True Story of a Lone Genius Who Solved the Greatest Problem of His Time* (New York: Penguin, 1996), 167–68.

12. Brown, *Story of Maps*, 99.

13. Alan Trachtenberg, *The Incorporation of America: Culture and Society in the Gilded Age* (New York: Hill and Wang, 1982), 19–20. More generally, see J. B. Harley, "Maps, Knowledge, and Power," in *The Iconography of Landscape: Essays on the Symbolic Representation, Design, and Use of Past Environments,* ed. Denis Cosgrove and Stephen Daniels (Cambridge: Cambridge University Press, 1988), 277–312.

14. Sobel, *Longitude,* 4–5.

15. Lewis Mumford, *Technics and Civilization* (New York: Harcourt, 1963), 13–16; David S. Landes, *Revolution in Time: Clocks and the Making of the Modern World* (Cambridge: Harvard University Press, 1983), 53–84.

16. Brown, *Story of Maps,* 230–40; Landes, *Revolution in Time,* 145–57; Sobel, *Longitude,* 106.

17. Sobel, *Longitude,* 11–20.

18. Landes, *Revolution in Time,* 151–55; Sobel, *Longitude,* 23–24.

19. Sobel, *Longitude,* 19. See also Landes, *Revolution in Time,* 155.

20. Sobel, *Longitude,* 134.

21. Landes, *Revolution in Time,* 77–78.

22. Albrecht Dürer, *Underweysung der Messung, The Printed Sources of Western Art,* ed. Theodore Besterman (Portland, OR: Collegium Graphicum, 1972); Eberhard Schröder, *Dürer: Kunst und Geometrie* (Basle: Birkhäuser, 1980).

23. Diringer, *Writing,* 62–63; I. J. Gelb, *A Study of Writing* (Chicago: University of Chicago Press, 1952), 155–57.

24. Diringer, *Writing,* 149–50; Gelb, *Study of Writing,* 178.

25. L. D. Reynolds and N. G. Wilson, *Scribes and Scholars: A Guide to the Transmission of Greek and Latin Literature* (Oxford: Clarendon, 1974), 30–32.

26. Ivan Illich, *In the Vineyard of the Text: A Commentary on Hugh's Didascalion* (Chicago: University of Chicago Press, 1993), 99–111.

27. Rosalind Thomas, *Oral Tradition and Written Record in Classical Athens* (Cambridge: Cambridge University Press, 1989), 95–154.

28. Elizabeth L. Eisenstein, *The Printing Revolution in Early Modern Europe* (Cambridge: Cambridge University Press, 1983), 7, 74, 201–2; Reynolds and Wilson, *Scribes and Scholars,* 200–12.

29. Brown, *Story of Maps,* 154–55.

30. Eisenstein, *Printing Revolution,* 197–200.

31. Ibid., 216–25.

32. Doron D. Swade, "Redeeming Charles Babbage's Mechanical Computer," *Scientific American,* February 1993, 86.

33. Aristotle, *Nicomachean Ethics* 5.5. See also Richard W. Hadden, *On the Shoulders of Merchants: Exchange and the Mathematical Conception of Nature in Early Modern Europe* (Albany: State University of New York Press, 1994), 83–156.

Chapter Eight: Realizing Information: Reading

1. D. C. Mitchell, *The Process of Reading: A Cognitive Analysis of Fluent Reading and Learning to Read* (Chichester: Wiley, 1982), 6–7.

2. Frank R. Vellutino, "Dyslexia," *Scientific American,* March 1987, 34–41.

3. Mitchell, *The Process of Reading,* 190–95.

4. Even in silent reading, the phonetic aspect of writing can swamp the logographic.

This is illustrated by *Ladle Rat Rotten Hut* (Amherst, MA: Nocturnal Canary Press, 1979) where every (phonetically and logographically correct) word of Little Red Riding Hood is replaced by one that is logographically wrong but phonetically close and the story remains intelligible. On the origin of *Ladle Rat Rotten Hut*, see *CoEvolution Quarterly* 18 (summer 1978): 138.

5. Paul Bertelson, "The Onset of Literacy," in *The Onset of Literacy: Cognitive Processes in Reading Acquisition*, ed. Bertelson (Cambridge: MIT Press, 1987), 8–9.

6. Jeanne S. Chall, *Stages of Reading Development* (New York: McGraw, 1983), 13–24; Bertelson, "The Onset of Literacy," 22–26.

7. Steven Pinker, *The Language Instinct* (New York: Morrow, 1994), 209.

8. Wolfgang Iser, *The Act of Reading: A Theory of Aesthetic Response* (Baltimore: Johns Hopkins University Press, 1978), 53–85.

9. Hugh of St. Victor, *Didascalion*, tr. and ed. Jerome Taylor (New York: Columbia University Press, 1961), 121.

10. Plato, *Phaedrus* 228D–E.

11. Plato, *Phaedrus* 275B–C, 278B–C, 279B.

12. The proper wording, spelling, and punctuation of James Joyce's *Ulysses* are famously and hotly contested issues. For the most recent dispute, see Sarah Lyall, "'Ulysses' in Deep Trouble Again," *New York Times,* 23 June 1997.

13. Norman Maclean, "A River Runs Through It," in *A River Runs Through It* (Chicago: University of Chicago Press, 1976), 1–104.

14. Hugh of St. Victor, *Didascalion*, 138. See also Ivan Illich, *In the Vineyard of the Text: A Commentary to Hugh's Didascalion* (Chicago: University of Chicago Press, 1993), 45–50.

15. In fact, the Septuagint was gradually compiled from the early third century B.C.E. through the early second century C.E.

16. Erich Schön, *Der Verlust der Sinnlichkeit oder die Verwandlungen des Lesers* (Stuttgart: Clett-Cotta, 1987), 99–122.

17. Augustine, *Confessions* 6.3 (my translation).

18. Paul Saenger, "Silent Reading: Its Impact on Late Medieval Script and Society," *Viator* 13 (1982), 367–414.

19. M. T. Clanchy, *From Memory to Written Record: England 1066–1307* (Cambridge: Harvard University Press, 1979), 215.

20. Illich, *In the Vineyard*, 60–61.

21. Ron McFarland and Hugh Nichols, eds., *Norman Maclean* (Lewiston, Idaho: Confluence Press, 1988), 84.

22. McFarland and Nichols, eds., *Norman Maclean*, 22.

23. Alan Cheuse, "Writing It Down for James: Some Thoughts on Reading Towards the Millennium," *Antioch Review* 51 (1993): 496–97.

24. For pictures of readers from the fifteenth to the nineteenth century, see Schön, *Der Verlust der Sinnlichkeit*. For reading contraptions, see the reading wheel (*Leserad*) of 1588 in Schön, fig. 3; the reader's window, advertised in the *New York Review of Books,* 12 December 1995; and the reader's table, advertised by Levenger (Delray Beach, Fla.).

25. Sven Birkerts, *The Gutenberg Elegies: The Fate of Reading in the Electronic Age* (New York: Fawcett, 1994), 77–86.

26. Dominik von König, "Lesesucht und Lesewut," in *Buch und Leser,* ed. Herbert G. Göpfert (Hamburg: Hauswedell, 1977), 62–75; Gerhard Sauder, "Gefahren empfind-

samer Vollkommenheit für Leserinnen und die Furcht vor Romanen in einer Damen-
bibliothek," in *Leser und Lesen im 18. Jahrhundert* (Heidelberg: Carl Winter, 1977),
83–91.

27. Chall, *Stages of Reading Development*.

28. For a richer account of the virtues and pleasures of reading, see Birkerts, *The
Gutenberg Elegies*, 11–113.

29. Cheuse, "Writing It Down for James," 486–87.

30. Irwin S. Kirsch, Ann Jungeblut, Lynn Jenkins, and Andrew Kolstad, *Adult Literacy in America*, 2d ed. (Washington, DC: Government Printing Office, 1993), xiv–xv.

31. Kirsch et al., *Adult Literacy*, xv.

32. Mild alarm and a primarily economic (rather than civic and cultural) orientation
are shown by the authors of *Adult Literacy*. Lawrence C. Stedman and Carl F. Kaestle
rightly call for "strenuous efforts to improve basic literacy skills" in Carl F. Kaestle,
Helen Damon-Moore, Lawrence C. Stedman, Katharine Tinsley, and William Vance
Trollinger Jr., *Literacy in the United States: Readers and Reading since 1880* (New Haven,
Conn.: Yale University Press, 1991), 127.

Chapter Nine: Realizing Information: Playing

1. Augustine, *Confessions* 8.12 (my translation).

2. Johann Sebastian Bach, "Meine Seel erhebt den Herren," *Die Bach Kantate*, vol.
17, conducted by Helmuth Rilling, Hänssler Classic compact disc 98.868; Johann Sebastian Bach, "Meine Seel erhebt den Herren," *Das Kantatenwerk*, vol. 3, conducted by
Gustav Leonhardt, Teldec compact disc 8.35029-1; Werner Neumann, *Handbuch der
Kantaten Johann Sebastian Bachs* (Leipzig: Breitkopf, 1971), 36.

3. Nelson Goodman, *Languages of Art: An Approach to a Theory of Symbols* (Indianapolis: Hackett, 1976), 187. See also p. 120, note 13.

4. Christoph Wolff, "Bach's Personal Copy of the Schübler Chorales" (first published in German in 1977), in *Bach* (Cambridge: Harvard University Press, 1991), 186.
For a critique of Wolff's account, see John Butt, *Bach Interpretation: Articulation Marks
in Primary Sources of J. S. Bach* (Cambridge: Cambridge University Press, 1990), 143.

5. Peter Kivy, *The Fine Art of Repetition: Essays in the Philosophy of Music* (Cambridge: Cambridge University Press, 1993), 35–74.

6. Johann Sebastian Bach, *Cantata Autographs in American Collections*, ed. Robert L.
Marshall (New York: Garland, 1985), ix.

7. According to Alfred Dürr, it is the ninth psalm tone. (Lutheran Church music
had nine, in place of the traditional eight, psalm tones.) See Dürr, *Die Kantaten von
Johann Sebastian Bach mit ihren Texten*, 6th ed. (Kassel: Bärenreiter, 1995), 750–51, 1025.

8. Fred Lerdahl and Ray Jackendoff, *A Generative Theory of Tonal Music* (Cambridge: MIT Press, 1983).

9. Hermann Hesse, *Das Glasperlenspiel*, Gesammelte Werke, vol. 9 (Frankfurt: Suhrkamp, 1970), 40–41 (my translation).

10. William Gibson, *Neuromancer* (New York: Ace Books, 1984), 5.

11. Gibson, *Neuromancer*, 5.

12. Natural numbers underdetermine not only their notation but also their structure
(as captured by axiomatization). See Gordon G. Brittan Jr., "The Ultimate Nature of
Physical Reality" (lecture presented at Duke University, 26 March 1991).

13. Donald Davidson, *Inquiries into Truth and Interpretation* (Oxford: Clarendon, 1984).

14. Jerrold Levinson, *Music, Art, and Metaphysics: Essays in Philosophical Aesthetics* (Ithaca, N.Y.: Cornell University Press, 1990), 63–88, 215–63.

15. Richard Taruskin, *Text and Act: Essays on Music Performance* (New York: Oxford University Press, 1995), 51–197.

16. Friedrich Blume, "Outline of a New Picture of Bach," *Music and Letters* 44 (1993): 220.

17. Quoted by Günther Stiller in *Johann Sebastian Bach and Liturgical Life in Leipzig,* tr. Herbert J. A. Bouman, Daniel F. Poellot, and Hilton C. Oswald (St. Louis: Concordia, 1984), 208. Stiller's entire book is devoted to supporting the claim that this testimony in fact reflects Bach's conception of music.

18. Luke 1:39–56.

19. As the *New Oxford Annotated Bible* (New York: Oxford University Press, 1973) has it on page 332. Hannah's song is in 1 Samuel 2:1–10.

20. Luke 1:5–25.

21. The translation is from the liner notes to *Die Bach Kantate.*

22. On the importance of the Hebrew scriptures in Bach's Cantatas, see Helene Werthemann, *Die Bedeutung der alttestamentlichen Historien in Johann Sebastian Bachs Kantaten* (Tübingen: Mohr, 1960). On Abraham in particular, see pp. 71–86.

23. Taruskin, *Text and Act,* 314.

24. Ibid., 309–10.

25. Ibid., 314.

Chapter Ten: Realizing Information: Building

1. In Daniel Dennett's *Darwin's Dangerous Idea: Evolution and the Meanings of Life* (New York: Simon and Schuster, 1995), this theme is at issue throughout under the headings of contingency, randomness, chance, and luck. Richard Rorty treats contingency explicitly in the first three chapters of *Contingency, Irony, and Solidarity* (New York: Cambridge University Press, 1989), 3–69. So does, in the context of evolution, Stephen Jay Gould in *Wonderful Life: The Burgess Shale and the Nature of History* (New York: Norton, 1989), 277–91. Steven Weinberg considers contingency under the heading of "historical accidents" in *Dreams of a Final Theory* (New York: Pantheon, 1992), 32–39, 164–165.

2. Contingency can be difficult and tends to divide people. Gould is inclined to find meaning in contingency while Dennett emphatically is not. On their disagreement see Dennett, *Darwin's Dangerous Idea,* pp. 262–312, and Gould, "Darwinian Fundamentalism" and "Evolution: The Pleasures of Pluralism," *New York Review of Books,* 12 June 1997, pp. 34–37, and 26 June 1997, pp. 47–52. Rorty himself is divided. A comparison of the earlier version of "The Contingency of Language" in *London Review of Books,* 17 April 1986, p. 3, with the later version in *Contingency, Irony, and Solidarity,* pp. 5 and 7, shows that in one place Rorty muted the voice of contingency and gave it more resonance in another.

3. The (relatively small) transept is omitted in this description but is discussed below.

4. Victor Hugo, *Le Rhin*, vol. 2, *Édition Nationale*, ed. Émile Testard, vol. 43 (Paris: Librairie de l'Édition Nationale, 1895), 198 (my translation).

5. Charles M. Radding and William W. Clark, *Medieval Architecture, Medieval Learning: Builders and Masters in an Age of Romanesque and Gothic* (New Haven, Conn.: Yale University Press, 1992).

6. Ernst Adam, *Das Freiburger Münster* (Stuttgart: Müller und Schindler, 1968) and "Das Bauwerk" in *Freiburger Münster*, ed. by the publisher (Freiburg: Rombach, 1982), 33–55.

7. Ernst Adam, "Der Turm des Freiburger Münsters," *Schau-ins-Land* 73 (1955), 21.

8. Walter Horn and Ernest Born, *The Plan of St. Gall*, 3 vols. (Berkeley and Los Angeles: University of California Press, 1979); Hans R. Hahnloser, *Villard de Honnecourt: Kritische Gesamtausgabe des Bauhüttenbuches ms. 19093 der Pariser Nationalbibliothek*, 2d ed. (Graz, Austria: Akademische Druck- und Verlagsanstalt, 1972).

9. A model can be found in the Stadtmuseum in Freiburg. For architectural drawings, see Adam, "Das Bauwerk," 36–37.

10. *The Papers of Thomas Jefferson*, vol. 13, ed. Julian P. Boyd (Princeton, N.J.: Princeton University Press, 1956), 267. See also p. 48 and the illustrations between pp. 16 and 17. On the connections between Strasbourg and Freiburg, see Adam, "Das Bauwerk," 46.

11. This aspect of gothic construction has been stressed by David Turnbull, "The Ad Hoc Collective Work of Building Gothic Cathedrals with Templates, String, and Geometry," *Science, Technology, and Human Values* 18 (1993): 315–40.

12. Hugo, *Le Rhin*, vol. 2, 200 (my translation). Hugo was right about the interplay of the Romanesque southern transept and the Renaissance portico, but wrong in thinking that the entire church was built from bottom (Romanesque) to top (Gothic). In fact it was built essentially from the middle (Romanesque transept and crossing) westward (Gothic nave and tower) and then eastward (late Gothic choir).

13. Matthew 18:19. This is how Caecilius Cyprianus (bishop of Carthage 249–58) translates the Greek *genēsetai* in "De Catholicae ecclesiae unitate," *Opera Omnia*, ed. Wilhelm Hartel (Vienna: C. Geroldi Filius Bibliopola Academiae, 1868), 220. The Vulgate has *fiet*.

14. For a historical and stylistic analysis of the tower, see Adam, "Der Turm des Freiburger Münsters," 18–65.

15. Otto von Simson, *The Gothic Cathedral: Origins of Gothic Architecture and the Medieval Concept of Order*, 2d ed. (Princeton, N.J.: Princeton University Press, 1962), 108–10.

16. Adam, "Der Turm des Freiburger Münsters," 36–45. Freiburg Minster became a cathedral (the seat of a bishop) in 1827.

17. Friedrich Ohly, "Die Kathedrale als Zeitenraum," *Schriften zur mittelalterlichen Bedeutungsforschung* (Darmstadt: Wissenschaftliche Buchgesellschaft, 1977), 171–91.

18. Victor Hugo, *Notre-Dame de Paris*, tr. Isabel F. Hapgood, 2 vols. (New York: Crowell 1888), 1:192.

19. On the medieval context of the *Synagoga-Ecclesia* relation, see Ohly, "Synagoge und Ecclesia," *Schriften zur mittelalterlichen Bedeutungsforschung*, 312–37.

20. Hugo, *Notre-Dame*, 1:189–90.

21. Simson, *The Gothic Cathedral*, 21–58.

22. An exhaustive discussion and refutation of proportionalism with the tower of Freiburg Minster as the principal instance has been provided by Konrad Hecht, "Mass und Zahl in der gotischen Baukunst," *Abhandlungen der Braunschweigischen Wissenschaftlichen Gesellschaft* 21 (1969): 215–326; 22 (1970): 105–263; 23 (1971–72): 25–236. On the tower of Freiburg Minster particularly, see vol. 21 (1969), pp. 274–319.

23. Villard de Honnecourt, *The Sketchbook of Villard de Honnecourt*, ed. Theodore Bowie (Westport, Conn.: Greenwood, 1982), pl. 35–38, 64.

24. Hecht's work has not discouraged latter-day proportionalists. An incredibly detailed and fanciful proportional proposal has been made by Friedrich Vellguth, *Der Turm des Freiburger Münsters* (Tübingen: Wasmuth, 1983). Appreciation of the accidental sense of contingency and infatuation with proportionalism can be found side by side in John James, *The Master Masons of Chartres* (Sydney: West Grinstead, 1990), 9–22, 83–128.

25. Reinhold Schneider, *Lyrik*, ed. Christoph Perels (Frankfurt: Insel, 1981), 158.

26. Peter Schickl, "Von Schutz und Autonomie zu Verbrennung und Vertreibung: Juden in Freiburg," *Geschichte der Stadt Freiburg*, ed. Heiko Haumann and Hans Schadek, vol. 1 (Stuttgart: Theiss, 1996), 524–34.

27. For a liberal account of these atrocities, see Heinrich Schreiber, *Geschichte der Stadt und Universität Freiburg im Breisgau* (Freiburg: Wangler, 1857), 136–41; for an antisemitic version, see Joseph Bader, *Geschichte der Stadt Freiburg im Breisgau* (Freiburg: Herder, 1883), 259–75.

28. Berent Schwineköper and Franz Laubenberger, "Geschichte und Schicksal der Freiburger Juden," *Freiburger Stadthefte* 4 (1963): 1–15; Schickl, "Von Schutz und Autonomie," 534–51.

29. Heiko Haumann, "Das Schicksal der Juden," *Geschichte der Stadt Freiburg im Breisgau*, ed. Haumann and Hans Schadek, vol. 3 (Stuttgart: Theiss, 1992), 325–39.

30. Reinhold Schneider, *Verhüllter Tag* (Frankfurt: Suhrkamp, 1980), 117 (my translation).

31. Charles Taylor, *Sources of the Self: The Making of the Modern Identity* (Cambridge: Harvard University Press, 1989), 491–92.

32. Karl Borgmann, "Vollbringer des Wortes," *Caritas* 59 (March/April 1958): nos. 3–4, pp. 87–89.

33. Schneider, *Lyrik*, 158 (my translation).

34. Gerd R. Ueberschär, "Die Stadt als Heimatfront im Zweiten Weltkrieg," *Geschichte der Stadt*, 3:361.

35. Gerd R. Ueberschär, *Freiburg im Luftkrieg 1939–1945* (Freiburg: Ploetz, 1990), 15, 264–65.

36. Wolfgang Fiek, "Die Reinzeichnungen aus dem Hubschrauber," *Badische Zeitung*, 30 May 1995.

Chapter Eleven: Elementary Measures

1. Plato, *Timaeus* 53A–56B.

2. Plato, *Timaeus* 80C. This is, with one revision, the R. G. Bury translation in the Loeb Classical Library (Cambridge: Harvard University Press, 1952).

3. Bern Dibner, "The Beginning of Electricity," *Technology in Western Civilization*,

ed. Melvin Kranzberg and Carroll W. Pursell Jr. (New York: Oxford University Press, 1967), 1:437–52; Steven Weinberg, *The Discovery of Subatomic Particles* (New York: Scientific American Books, 1983), 14–20.

4. Assuming units of newton, kilogram, meter, and second.

5. Weinberg, *Discovery*, 20–71.

6. Ibid., 172–73.

7. John Haugeland, "Semantic Engines," *Mind Design: Philosophy, Science, and Artificial Intelligence*, ed. Haugeland (Cambridge: MIT Press, 1981), 23.

8. Anton Glaser, *History of Binary and Other Nondecimal Numeration* (Los Angeles: Tomash, 1981), 31–37.

9. George A. Miller, "The Magical Number Seven, Plus or Minus Two," *Psychological Review* 63 (1956): 81–97.

10. The notion was coined in 1947 by J. W. Tukey. See H. S. Tropp, "The Origin of the Term Bit," *Annals of the History of Computing* 6 (1984): 152–55.

11. Claude E. Shannon, "The Mathematical Theory of Communication" (first published in 1948), in Shannon and Warren Weaver, *The Mathematical Theory of Communication* (Urbana: University of Illinois Press, 1949), 3–91.

12. Warren Weaver, "Recent Contributions to the Mathematical Theory of Communication," in Shannon and Weaver, *Mathematical Theory*, 114–15.

13. Yehoshua Bar-Hillel, *Language and Information: Selected Essays on Their Theory and Application* (Reading, Mass.: Addison-Wesley, 1964), 292. (This is from an essay first published in 1955.)

14. Donald M. MacKay, *Information, Mechanism, and Meaning* (Cambridge: MIT Press, 1969), 17.

15. R. V. L. Hartley, "Transmission of Information," *Bell System Technical Journal* 7 (1928): 536.

16. MacKay, *Information*, 17.

17. See Hartley, "Transmission," pp. 536–40, on the technical reasons for taking "as our practical measure of information the logarithm of the number of possible symbol sequences" (p. 540). See also Shannon, "Mathematical Theory," 3–4. On some of the advantages of the base-two logarithm, see Doede Nauta Jr., *The Meaning of Information* (The Hague: Mouton, 1972), 182–83.

18. There is, however, another and perhaps more persistent ambiguity in our belief that the amount of information corresponds to the unpredictability of information. The simplest way of capturing this intuition is not to use a logarithmic function but to define it as the complement of the probability of information to one. Thus for the three instances of John and Abe's situation the amounts of information are (1) $1 - \frac{1}{2} = .5$, (2) $1 - \frac{1}{4} = .75$, (3) $1 - \frac{1}{8} = .875$. For a detailed exposition and discussion, see Bar-Hillel, *Language*, 221–74, 298–310. See also Nauta, *Meaning of Information*, 214–20.

19. Nauta, *Meaning*, 179–84.

20. The closest, if most artificial, match between information and reality is reached when things are so partitioned into possible states and signs are so structured that the number of (states converted via the logarithm to base 2 into) bits equals the number of signs. This is true of John's and Abe's case and, in a very different way, of virtual realities.

21. Bar-Hillel, *Language*, 226–27, 300–301.

22. Ibid., 284, 298.

23. Shannon, "The Mathematical Theory," 3. On the more flagrant confusion of Hartley, "Transmission," p. 536 (the passage quoted above), see Bar-Hillel, *Language,* 283–86.

24. Weaver, "Recent Contributions," 100.

25. J. Murray Barbour, "Bach and *The Art of Temperament,*" *Musical Quarterly* 33 (1947): 64–89. Herbert Kelletat, *Zur musikalischen Temperatur* (Kassel: Merseburger, 1981), 1:21–25, 52–54.

Chapter Twelve: Basic Structures

1. Gen. 1:1–4.

2. Anton Glaser, *History of Binary and Other Nondecimal Numeration* (Los Angeles: Tomash, 1981), 31.

3. Ibid., 32.

4. Ferdinand de Saussure, *Course in General Linguistics,* ed. Charles Bally and Albert Sechehaye, tr. Wade Baskin (1915; reprint, New York: McGraw-Hill, 1966), 120.

5. Martin Heidegger, *Identität und Differenz* (Pfullingen: Neske, 1957).

6. John Archibald Wheeler, "Information, Physics, Quantum," *Complexity, Entropy, and the Physics of Information,* ed. Wojciech H. Zurek (Redwood City, Cal.: Addison-Wesley, 1990), 5.

7. Doron D. Swade, "Redeeming Charles Babbage's Mechanical Computer," *Scientific American,* February 1993, 91.

8. Herman H. Goldstine, *The Computer from Pascal to Von Neumann* (Princeton, N.J.: Princeton University Press, 1993), 184.

9. Goldstine, *The Computer,* 231.

10. Robert N. Noyce, "Microelectronics," *Scientific American,* September 1977, 65.

11. Richard Coyne, *Designing Information Technology in the Postmodern Age: From Method to Metaphor* (Cambridge: MIT Press, 1995), 308.

12. Analogously, conjunctions and disjunctions can be used to reflect how much information we have about what is the case in reality. This notion has been rendered precise for an extremely limited example by Yehoshua Bar-Hillel in *Language and Information: Selected Essays on Their Theory and Application* (Reading, Mass.: Addison-Wesley, 1964), 221–48.

13. The first to publish a demonstration that one binary operation is sufficient was Edward V. Huntington, "Sets of Independent Postulates for the Algebra of Logic," *Transactions of the American Mathematical Society* 5 (1904): 288–309. The first to make the application to propositional logic was Henry Maurice Sheffer, "Postulates for Boolean Algebra," *Transactions* 14 (1913): 486–88.

14. Actually, one of two possible operations suffices. See Willard Van Orman Quine, *Mathematical Logic* (New York: Harper, 1962), 45–49.

15. Intuitively it is not obvious why a juncture should correspond to AND rather than OR. In fact the latter seems more plausible. But due to the contingencies of transistors, a juncture is biased toward a voltage drop (just as AND is biased toward FALSE), low voltage being conventionally associated with 0 or FALSE.

16. Gordon G. Brittan Jr., "Constructibility and the World-Picture," *Proceedings: Sixth International Kant Congress,* ed. G. Funke and Th. M. Seebohm (Washington, D.C.: University Press of America, 1989), 65.

17. Bill Gates, *The Road Ahead* (New York: Viking, 1995), xi, 14–17.

18. That is two million for the Intel 8080, 400 million for Intel's Pentium II.

19. H. M. Deitel and P. J. Deitel have likened the computer to a (more up-to-date) manufacturing plant in *C How to Program* (Englewood Cliffs, N.J.: Prentice Hall, 1992), 4.

20. Edgar Allen Poe was intrigued with Maelzel's automaton and unmasked it in "Maelzel's Chess-Player" (first published in 1836), *Essays and Reviews* (New York: Library of America, 1984), 1253–76.

21. Edward Roberts and William Yates, "Altair 8800: The Most Powerful Minicomputer Project Ever Presented—Can Be Built for Under $400," *Popular Electronics*, January 1975, 31–38.

22. Edward Roberts and William Yates, "Build the Altair 8800 Minicomputer: Part Two," *Popular Electronics*, February 1975, 56.

23. Sherry Turkle, *The Second Self: Computers and the Human Spirit* (New York: Simon and Schuster, 1984), 165–95.

24. See Steven Levy's chapter on "The Hacker Ethic" in his *Hackers: Heroes of the Computer Revolution* (Garden City, N.Y.: Anchor, 1984), 26–36.

25. No one has embodied this vision as well as Stewart Brand who distinguished himself not only as the founder, publisher, and editor of the *Whole Earth Catalog* (1968–85) and the *CoEvolution Quarterly* (1973–84), but also as the author of *The Media Lab: Inventing the Future at MIT* (New York: Viking, 1987).

26. Noyce, "Microelectronics," 65.

Chapter Thirteen: Transparency and Control

1. Alvin F. Harlow, *Old Wires and New Waves: The History of the Telegraph, Telephone, and Wireless* (New York: Appleton-Century, 1936); Herman H. Goldstine, *The Computer: From Pascal to Von Neumann* (Princeton, N.J.: Princeton University Press, 1993).

2. In addition to the bit and the tit, there is the dit, corresponding to the decimal base, and the nit, corresponding to base e (of the natural logarithm). See Doede Nauta Jr., *The Meaning of Information* (The Hague: Mouton, 1972), 183 n. 14.

3. Robert R. Birge, "Protein-Based Computers," *Scientific American*, March 1995, 90–96; Jim Ludwick, "Missoula Man May Shed Light on Transistors," *Missoulian*, 13 March 1994; Seth Lloyd, "Quantum-Mechanical Computers," *Scientific American*, October 1995, 140–45.

4. Frances Yates, *The Art of Memory* (Chicago: University of Chicago Press, 1966), xi. See also p. 4.

5. Yates, *Art of Memory*, 7.

6. Meriwether Lewis and William Clark, *The History of the Lewis and Clark Expedition*, ed. Elliott Coues, 3 vols. (1893; reprint, New York: Dover, n.d.), 1:221.

7. Lewis and Clark, *History*, 2:802–3.

8. Ibid., 3:1071.

9. Edward R. Tufte, *Envisioning Information* (Cheshire, Conn.: Graphics, 1990), 80.

10. Thomas Jefferson, "Instructions to Captain Lewis," *The Portable Thomas Jefferson*, ed. Merrill D. Peterson (Harmondsworth: Penguin, 1975), 309.

11. Ibid., 310.

12. Arlen J. Large, "Lewis and Clark: Part Time Astronomers," *We Proceeded On* 5 (February 1979): 9. See also Stephen E. Ambrose, *Undaunted Courage* (New York: Simon and Schuster, 1996), 87.

13. Large, "Lewis and Clark," 9–10; Ambrose, *Undaunted Courage*, 119.

14. Large, "Lewis and Clark," 10–11.

15. Edward R. Tufte, *The Visual Display of Quantitative Information* (Cheshire, Conn.: Graphics, 1983), 162, and *Envisioning*, 81.

16. J. T. Coppock and D. W. Rhind, "The History of GIS," *Geographical Information Systems*, ed. David J. Maguire, Michael F. Goodchild, and David W. Rhind (Harlow: Longman, 1991), 21–43.

17. M. F. Goodchild, "The Technological Setting of GIS," *Geographical Information Systems*, 45–54.

18. Gordon G. Brittan Jr., "Constructibility and the World-Picture," *Proceedings: Sixth International Kant Congress*, ed. G. Funke and Th. M. Seebohm (Washington, D.C.: University Press of America, 1989), 65.

19. A. C. Gatrell, "Concepts of Space and Geographical Data," *Geographical Information Systems*, 119–34.

20. On the layers of information, see Tufte, *Envisioning*, 65.

21. *Northwest Montana from Space*. (Missoula, Mont.: Wildlife Spatial Analysis Lab, n.d.). Available: http://www.wru.umt.edu/posters/200/tm.missoula.html. 10 June 1998.

22. Diane Wickland, "Mission to Planet Earth," *Ecology* 72 (1991): 1928–31.

23. *Where We Live in the Greater Missoula Area*. (Missoula, Mont.: Wildlife Spatial Analysis Lab, n.d.). Available: http://www.wru.umt.edu/posters/200/greater.msla. ppn.html. 10 June 1998.

24. Walter Gekelman, James Bamber, David Leneman, Steve Vincena, James Maggs, and Steve Rosenberg, "Making Sense from Space-Time Data in Laboratory Experiments on Space Plasma Processes," *Visualization Techniques in Space and Atmospheric Sciences*, ed. E. P. Szuszczewicz and J. H. Bredekamp (NASA 1.21:519), 221–32.

25. Hassan Aref, Richard D. Charles, and T. Todd Elvins, "Scientific Visualization of Fluid Flow," *Frontiers of Scientific Visualization*, ed. Clifford A. Pickover and Stuart K. Tewksbury (New York: Wiley, 1994), 7–43.

26. The answer is yes. See John D. Barrow, *Pi in the Sky: Counting, Thinking, and Being* (Boston: Little, Brown, 1992), 227–34. Another problem that has been solved with the help of computers is the version of the party problem where you inevitably have four friends or five strangers at the party (and it turns out that you need at least 25 guests). See John Horgan, "The Death of Proof," *Scientific American*, October 1993, 100, 101.

27. Horgan, "Death of Proof," 95–96.

28. On Life, see Rudy Rucker, *Exploring Cellular Automata* (Sausalito, Cal.: Autodesk, 1989), 199–203; and Daniel Dennett, *Darwin's Dangerous Idea: Evolution and the Meanings of Life* (New York: Simon and Schuster, 1995), 166–76. See also Philippe de Reyffe, "Computer Simulation of Plant Growth," *Frontiers*, 145–79.

29. On the reservations of mathematicians, see Horgan, "Death of Proof," pp. 92–102 and Barrow, *Pi in the Sky*, pp. 227–43.

30. Sherry Turkle, *Life on the Screen: Identity in the Age of the Internet* (New York: Simon and Schuster, 1995), 160, 167.

31. Viglius Zuichemus writing to Erasmus, in *Opus epistolarum Des. Erasmi Roterdami: Complete Letters of Erasmus,* 12 vols., ed. P. S. Allen (Oxford: Oxford University Press, 1992), 10: 29–30.

32. Bill Gates, *The Road Ahead* (New York: Viking, 1995), 79–88. On time as an alternative dimension for ordering information, see Jeffrey R. Young, "New Metaphors for Organizing Data Could Change the Nature of Computers," *Chronicle of Higher Education,* 4 April 1997, A19–A20.

33. Tufte, *Visual Display,* 191. "Information Architects" have taken it upon themselves, sometimes in defiance of Tufte's principles of clarity and economy, to control and organize information. See Richard Saul Wurman and Peter Bradford, eds., *Information Architects* (Zurich: Graphics, 1996).

34. B. P. Buttenfield and W. A. MacKaness, "Visualization," *Geographical Information Systems,* 436–37.

35. Melissa Marie Hart, "Past and Present Vegetative and Wildlife Diversity in Relation to an Existing Reserve Network" (master's thesis, University of Montana, 1994), 14–24.

36. Gatrell, "Concepts of Space," 128; J.-C. Muller, "Generalization of Spatial Databases," *Geographical Information Systems,* 458.

37. Muller, "Generalization," 457–75.

38. Horgan, *The End of Science: Facing the Limits of Knowledge in the Twilight of the Scientific Age* (Reading, Mass.: Addison-Wesley, 1996), 191–246.

39. Hubert L. Dreyfus, *What Computers Still Can't Do: A Critique of Artificial Reason,* rev. ed.(Cambridge: MIT Press, 1992). The first edition was published in 1972.

40. Goodchild, "The Technological Setting of GIS," 53.

41. Buttenfield and MacKaness, "Visualization," 437. Sherry Turkle has called the transparency of the substructure "modernist" and the transparency of the superstructure "postmodernist." See *Life,* 36–43.

42. Turkle, *Life,* 60. See also 63–66.

43. J. Roughgarden, S. W. Running, and P. A. Matson, "What Does Remote Sensing Do for Ecology?" *Ecology* 72 (1991), 1918–22; Hart, "Past and Present," 1–5.

44. See my *Technology and the Character of Contemporary Life: A Philosophical Inquiry* (Chicago: University of Chicago Press, 1984), 35–40.

45. See my *Crossing the Postmodern Divide* (Chicago: University of Chicago Press, 1992), 20–47.

46. Bill McKibben, *The Age of Missing Information* (New York: Penguin, 1993), 110.

47. Paul Proctor, "Boeing Rolls Out 777 to Tentative Market," *Aviation Weekly and Space Technology,* 11 April 1994, 36.

48. Karl Sabbagh, *21st-Century Jet: The Making and Marketing of the Boeing 777* (New York: Scribner, 1996), 56–67.

49. Proctor, "Boeing Rolls Out 777," 37.

50. Sabbagh, *21st-Century Jet,* 58; Proctor, "Boeing Rolls Out 777," 37.

Chapter Fourteen: Virtuality and Ambiguity

1. Claude E. Shannon, "The Mathematical Theory of Communication," in Shannon and Warren Weaver, *The Mathematical Theory of Communication* (Urbana: University of Illinois Press, 1949), 3. Shannon, unhelpfully and somewhat inconsistently, given his

diagram on p. 5 and his remarks on pp. 4–6, equates "message" and "structure of the signal." I am following Weaver who, on p. 100 of "Recent Contributions to the Mathematical Theory of Communication," not quite consistently either, identifies "message" with "meaning of the signal."

2. Weaver, "Recent Contributions," 100.

3. This is a rough estimate, using ASCII as the measure. Strictly speaking, the message here is itself a signal whose meaning can be realized through reading and living.

4. One who attacked this problem head-on was Donald McKay, who called resolution the "structural information content" and suggested the "logon" (specifying "one independent element or 'degree of freedom' of a signal") as the unit of measurement. The latter never caught on because it was intuitively too precise and technically not precise enough. See his *Information, Mechanism, and Meaning* (Cambridge: MIT Press, 1969), 1–18. For criticism, see Doede Nauta Jr., *The Meaning of Information* (The Hague: Mouton, 1972), 207–14.

5. Compare, e.g., Günther Ramin's recording of 1941 (Berlin: Electrola [now available on compact disc—Calig-Verlag 50859–60]) with Philippe Herreweghe, *J. S. Bach: Matthäus Passion* (Saint-Michel de Provence: Harmonia Mundi, 1985). This example, incidentally, shows that substance and resolution can be inversely related.

6. Generally, if resolution is low, the demands on intelligence are high. See Yehoshua Bar-Hillel, *Language and Information: Selected Essays on Their Theory and Application* (Reading, Mass.: Addison-Wesley, 1964), 277. If resolution is too low it overtaxes intelligence, if it is too high it exceeds legibility.

7. Again with ASCII as the measure.

8. Once more a rough estimate, using ASCII and assuming twelve bits per note.

9. Helmuth Rilling, *Die Bach Kantate*, vol. 17, Hänssler Classic 98.868.

10. A distinctive term for real reality needs to be found now that reality can be virtual. *Actual reality* is the obvious choice. When there is no danger of confusion I will use "reality" in place of "actual reality" and "real" in place of "actual."

11. Jonathan Steuer, "Defining Virtual Reality," *Journal of Communication* 42 (1992): 79–90.

12. For an introduction to the technology of virtual reality, see Frank Biocca, "Virtual Reality Technology: A Tutorial," *Journal of Communication* 42 (1992): 23–72. For the state of the art in 1991, and hopes for the future, see Howard Rheingold, *Virtual Reality* (New York: Oxford University Press, 1993). For philosophical reflections on virtual reality, see Michael Heim, *The Metaphysics of Virtual Reality* (New York: Oxford University Press, 1993).

13. Tony Verna in his contribution to a collection of prognostications gathered by Marion Long, "The Seers' Catalogue," *Omni* 9 (January 1987): 40.

14. The Bullfrog Community Hospital is one of the toy universes that allows for a precise measurement of information about reality. For "Paging Doctor Alphonso" it is two bits, for "Paging Doctor Alfie" it is one bit, for "Paging Doctor Al" it is zero bits. Though the example is misleading as regards precision, it does illustrate the broader and valid point about the inverse relation of ambiguity and information about reality.

15. For a discussion of the interplay of illusion and reality and of art and religion at the threshold of the modern era, see Karsten Harries, *The Bavarian Rococo Church: Between Faith and Aestheticism* (New Haven, Conn.: Yale University Press, 1983).

16. William J. Mitchell, *City of Bits: Space, Place, and the Infobahn* (Cambridge:

MIT Press, 1995), 167. Similar sentiments appear on pp. 20, 31, 36, 44, 107, and 152. But Mitchell has reservations too; see pp. 19, 44, 75, 98, 104, and 169–70.

17. Sherry Turkle, *Life on the Screen: Identity in the Age of the Internet* (New York: Simon and Schuster, 1995), 193.

18. David Bennahum, "Fly Me to the MOO," *Lingua Franca*, May/June 1994, 26.

19. See Turkle, *Life*, pp. 180–86, for a more detailed account.

20. For the elusiveness of artificial intelligence, see Hubert L. Dreyfus, *What Computers Still Can't Do: A Critique of Artificial Reason* (Cambridge: MIT Press, 1992).

21. Turkle, *Life*, 177–269.

22. See my "Artificial Intelligence and Human Personality," *Research in Philosophy and Technology* 14 (1994): 271–83.

23. Turkle, *Life*, 88–94.

24. Ibid., 207.

25. Bennahum, "Fly Me," 24.

26. For the extent of sex on the World Wide Web, see Matt Richtel, "Sex Sites Exxx-posed," *Internet Underground* 2 (May 1997): 28–33. Available: http://search.zdnet.com/iu/archive/issue17/sexsites/. 10 June 1998.

27. "A Death On-Line Shows a Cyberspace with Heart and Soul," *New York Times*, 23 April 1995.

28. Ibid.

29. Jon Katz, "The Tales They Tell in Cyber-Space Are A Whole Other Story," *New York Times*, 23 January 1994.

30. Peter Cortland, "Recycled Sob Stories," *New York Times*, 20 February 1994.

31. Jonathan Vankin and John Whalen, "How a Quack Becomes a Canard," *New York Times Magazine*, 17 November 1996, 56–57.

32. Deborah Tannen, "Gender Gap in Cyberspace," *Newsweek*, 16 May 1994, 52.

33. Alison L. Sprout, "Saving Time around the Clock," *Fortune*, 13 December 1993, 157.

34. For an example of how humor can dispel cyberspace fog, see Gloria Mitchell, "Next Time You Think, Think Fertnel," *Internet Underground* 2 (April 1997): 35–36. Available: http://search.zdnet.com/iu/archive/issue16/cheese/index.html. 10 June 1998.

Chapter Fifteen: Fragility and Noise

1. *The Works of Horace*, ed. E. C. Wickham (Oxford: Clarendon, 1877), 1:255.

2. My translation.

3. Charles A. Vannoy, *Studies on the Athena Parthenos of Pheidias* (Iowa City: University of Iowa, 1914). See pp. 54–55 in particular.

4. Jeff Rothenberg, "Ensuring the Longevity of Digital Documents," *Scientific American*, January 1995, 42.

5. Ibid., 44.

6. Avra Michelson and Jeff Rothenberg, "Scholarly Communication and Information Technology," *American Archivist* 55 (1992): 295–301.

7. Neil Munro, for one, thinks that "A wonk armed with a computer could bring America to its knees." See his "Our Electronic Achilles' Heel," *Washington Post Weekly Edition*, 14–20 August 1995, 24.

8. Ronald L. Enfield, "The Limits of Software Reliability," *Technology Review*, April 1987, 36–43; Bev Littlewood and Lorenzo Strigini, "The Risks of Software," in *Computers, Ethics, and Social Values*, ed. Deborah G. Johnson and Helen Nissenbaum (Englewood Cliffs, N.J.: Prentice Hall, 1995), 432–37.

9. Philip E. Ross, "The Day the Software Crashed," *Forbes*, 25 April 1994, 142–56; Nancy G. Leveson and Clark S. Turner, "An Investigation of the Therac-25 Accidents," in *Computers, Ethics, and Social Values*, ed. Johnson and Nissenbaum, 474–514.

10. Do the 65,536 numbers really constitute a notational system by, e.g., Nelson Goodman's definition? In a strained and qualified sense the answer is yes. The numbers satisfy the type-token distinction, given the parsing convention of sixteen bits. They are identifiable to a CD player though not unfailingly so, a flaw compensated for by redundancy. They are unambiguous in referring to one and only one component of the wave form though they are vastly ambiguous relative to an entire piece of music (but so is, if much less so, a single note). Their referents (the microsegments of the waveform) are theoretically distinct from one another. They fail Goodman's last criterion in that no two digital recording systems will segment a performance into the same sequence of (binary) numbers. But then would Mozart and Haydn on hearing Allegri's *Miserere* have come up with exactly the same score? See Goodman, *Languages of Art: An Approach to a Theory of Symbols* (Indianapolis: Hackett, 1976), 132–35, 143, 156.

11. Parvis Moin and John Kim, "Talking Turbulence with Supercomputers," *Scientific American*, January 1997, 62–68.

12. Robert M. Solow, "How Did Economics Get That Way and What Way Did It Get?" *Daedalus* 126 (winter 1997): 43–53.

13. Naomi Oreskes, Kristin Shrader-Frechette, Kenneth Belitz, "Verification, Validation, and Confirmation of Numerical Models in the Earth Sciences," *Science* 263 (4 February 1994): 641–46; Brian Cantwell Smith, "Limits of Correctness in Computers," in *Computers, Ethics, and Social Values*, ed. Johnson and Nissenbaum, 456–69.

14. Paul Wallich, "The Wall Falls," *Scientific American*, August 1996, 20.

15. W. Wayt Gibbs, "Virtual Reality Check," *Scientific American*, December 1994, 40–42.

16. William J. Mitchell, *City of Bits: Space, Place, and the Infobahn* (Cambridge: MIT Press, 1995), 106; Barbara Kantrowitz, "The Metaphor is the Message," *Newsweek*, 14 February 1994, 49.

17. For a portrayal of a virtual golf course, see Gary Jahrig, "Swing Training," *Missoulian*, 29 February 1996.

18. For reviews of the virtual gallery, see Bernard Sharratt, "Please Touch the Paintings," *New York Times Book Review*, 6 March 1994, 3, 18; and Ben Davis, "The Gallery in the Machine," *Scientific American*, May 1995, 107–10.

19. David Bennahum, "Fly Me to the MOO," *Lingua Franca*, May/June 1994, 26; Gen. 29:17.

20. Gen. 29:20.

21. Kenneth John Conant, "Medieval Academy Excavations at Cluny," *Speculum* 4 (1929): 3–4.

22. This is a somewhat unorthodox version of noise. I owe it to Fred I. Dretske, *Knowledge and the Flow of Information* (Cambridge: MIT Press, 1981), 20–21, 239 n 13.

23. Again I am following Dretske, *Knowledge*, 21.

24. D. T. Max, quoting Louis Rosetto, in "The End of the Book?" *Atlantic Monthly*,

September 1994, 62; John Perry Barlow in "What Are We Doing On-Line?" *Harpers*, August 1995, 36.

25. *Newsweek*, 30 June 1980, 50. See also very similar pronouncements by Michael Dertouzos in "Communications, Computers, and Networks," *Scientific American*, September 1991, 62; Gerald M. Levin in Ken Auletta, "The Magic Box," *New Yorker*, 11 April 1994, 41; and Rosetto in Max, "The End of the Book?" 62.

26. John James, *The Master Masons of Chartres* (Sydney: West Grinstead, 1990), 134–44.

27. Stephen C. Ehrmann, "Moving Beyond Campus-Bound Education," *Chronicle of Higher Education*, 7 July 1995, 2; Diana G. Oblinger and Mark K. Maruyama, "Distributed Learning," CAUSE Professional Paper Series, no. 14 (1996): 14.

28. Gen. 42. Oblinger and Maruyama invoke the more contemporary "factory model" where students presumably are the drudges and instructors the bosses. See "Distributed Learning," 3.

29. *Changing the Process of Teaching and Learning: Essays by Notre Dame Faculty* (Notre Dame, Ind.: University of Notre Dame, 1994).

30. Ibid., 20–21, 50, 70; Oblinger and Maruyama, "Distributed Learning," 13.

31. Oblinger and Maruyama, "Distributed Learning," 7–10.

32. For "having the world database at your fingertips," see Bill Gates's contribution to a collection of prognostications gathered by Marion Long in "The Seers' Catalog," *Omni*, January 1987, 38. As for a more explicit analogy of the computer person and the personal computer, the system memory, working memory, or random access memory (RAM) of the computer is empty when the computer is turned off. The information processing prowess of a computer is permanently stored in the central processing unit (CPU). The piece of software minimally required to boot up the computer, the basic input/output system (BIOS), is permanently stored in a read-only-memory (ROM) chip. The operating system is stored on the hard disk and loaded into the system memory when the computer is booted up. A lesser operating system (e.g., DOS) can be replaced by a better one (e.g., Windows 95).

33. E. D. Hirsch, *The Schools We Need and Why We Need Them* (New York: Doubleday, 1996), 69–126.

34. Neil L. Rudenstine, president of Harvard University, listed the following bullets in an article about the Internet:

- The Internet can provide access to essentially unlimited sources of information not conveniently obtainable through other means.
- The Internet allows for the creation of unusually rich course materials.
- The Internet enhances the vital process of "conversational" learning.
- The Internet reinforces the conception of students as active agents in the process of learning, not as passive recipients of knowledge from teachers and authoritative texts.

Chronicle of Higher Education, 21 February 1997, A48. Reed Browning, former provost of Kenyon College, describes the Internet as "an almost inexpressibly grand tool for learning and instruction" in "What Presidents Need to Know about the Payoff on the Information Investment: The View from Kenyon College." Available: http://www.kenyon.edu/dpts/ics/docs/policy/hierkc.htm.iic. 26 June 1998. Richard A.

Crofts, Commissioner of the Montana University System, has said, "For the first time in the history of American higher education, information technology provides the opportunity to increase access to higher education, improve the quality of students' learning experiences, enhance the faculty role as teacher/scholar/learner, and control the costs of education—simultaneously." See his "The Montana Learner Imperative." Available: http://www.montana.edu/www/bor/docs/learner.html. 26 June 1998. Others, however, think that information technology will not improve higher education as we know it but rather make it obsolete. See Lewis J. Perelman, *School's Out: Hyperlearning, the New Technology, and the End of Education* (New York: William Morrow, 1992), 19–24; Peter Drucker, quoted by Robert Lenzner and Stephen J. Johnson in "Seeing Things as They Really Are," *Forbes*, 10 March 1997, 127.

35. Robert L. Jacobson, "As Instructional Technology Proliferates, Skeptics Seek Hard Evidence of Its Value," *Chronicle of Higher Education*, 5 May 1993, A27–A29; Charles R. McClure and Cynthia L. Lopata, *Assessing the Academic Networked Environment: Strategies and Options* (Washington, D.C.: Association of Research Libraries, 1996), 1–3.

36. One early meta-analysis bravely reported "small but significant contributions" of computer-based college teaching and another a "moderate-size but statistically significant effect" of computer-based adult education. See James A. Kulik, Chen-Lin C. Kulik, and Peter A. Cohen, "Effectiveness of Computer-Based College Teaching," *Review of Educational Research* 50 (1980): 525–44; and Chen-Lin C. Kulik, James A. Kulik, and Barbara J. Schwalb, "The Effectiveness of Computer-Based Adult Education," *Journal of Educational Computing Research* 2 (1986): 235–52. On the doubtful benefits of computers for primary and secondary education, see Todd Oppenheimer, "The Computer Delusion," *Atlantic Monthly*, July 1997, 45–62.

37. Bruce V. Lewenstein, "Do Public Electronic Bulletin Boards Help Create Scientific Knowledge?" *Science, Technology, and Human Values* 20 (1995): 138.

38. Ibid., 140.

39. Ibid., 143.

40. For more on *Perseus* see http://www.perseus.tufts.edu (10 June 1998). For reservations about *Perseus*, see Sven Birkerts, *The Gutenberg Elegies: The Fate of Reading in an Electronic Age* (New York: Fawcett, 1994), 134–40.

41. Karen Ruhleder, "Reconstructing Artifacts, Reconstructing Work: From Textual Edition to On-Line Databank," *Science, Technology, and Human Values* 20 (1995): 49.

42. Ibid., 56–57.

43. Ibid., 51–56.

44. Jay David Bolter, *Writing Space: The Computer, Hypertext, and the History of Writing* (Hillsdale, N.J.: Erlbaum, 1991), 142.

45. Ibid., 25.

46. Robert Coover, "The End of Books," *New York Times Book Review*, 21 June 1992, 25. Elsewhere Coover has expressed more dismay yet and noted that the fiction and poetry created in Brown University's Hypertext Hotel leads, due to the lack of editorial control, to a reading experience that is "very diffuse, and seems totally undirected, and is finally kind of a bore to move around in." (Quoted by David Bennahum in "Fly Me to the MOO," *Lingua Franca*, May/June 1994, 31.) Similarly, Christopher L.

Tomlins has called the reviews that appeared in the scholarly region of cyberspace occupied by the H-Net (the history net) "long-winded, turgid, rambling, bereft of wit or flair." See his "Print and Electronic Book Reviewing Can Peacefully Co-exist," *Chronicle of Higher Education*, 9 August 1996, A40. For all its airiness, cyberspace seems to exert approval pressure on its critics, or else things are truly getting better. Tomlins thinks that "the quality of on-line reviews is improving" (p. A40), and Coover takes a rather hopeful and occasionally enthusiastic view of the new medium in "Hyperfiction: Novels for the Computer," *New York Times Book Review*, 29 August 1993, 1, 8–12.

47. Stuart Moulthrop, *Victory Garden: A Fiction* (Cambridge, Mass: Eastgate Systems, 1991). For a celebration of *Victory Garden* see Coover, "Hyperfiction," 1, 8–10, 11. For criticism, see Birkerts, *Gutenberg Elegies*, 151–64.

48. Thomas K. Landauer, *The Trouble with Computers: Usefulness, Usability, and Productivity* (Cambridge: MIT Press, 1995), 15.

49. W. Wayt Gibbs, "Taking Computers to Task," *Scientific American*, July 1997, 82–86.

50. For a German sampling of hopeful expectations see Peter Brödner, *Der überlistete Odysseus* (Berlin: Sigma, 1997), 17.

51. Jeff Madrick, "The Cost of Living: A New Myth," *New York Review of Books*, 6 March 1997, 19–24.

52. Landauer, *Trouble with Computers*, 9–77; Brödner, *Der überlistete Odysseus*, 15.

53. For an account of a typical case, see Landauer, *Trouble with Computers*, 1–3.

54. Ibid., 4, 166–67; Richard A. Spinello, *Case Studies in Information and Computer Ethics* (Upper Saddle River, N.J.: Prentice Hall, 1997), 203–5.

55. Mitch Betts, "IS Can Control Internet Surfing, Misuse," *Computer World*, 14 February 1994, 1, 26.

56. Sally Tisdale, "Silence, Please," *Harpers*, March 1997, 65–74.

Conclusion: Information and Reality

1. See, e.g., Charles Leroux, "Drowned in Data?" *Chicago Tribune*, 15 October 1996; Richard Zoglin, "The News Wars," *Newsweek*, 21 October 1996, 58–64; Richard Harwood, "40 Percent of Our Lives," *Washington Post*, 30 November 1996; Louis Uchitelle, "What Has the Computer Done for Us Lately?" *New York Times*, 8 December 1996; and the 16 December 1996 cover of the *New Yorker* contrasting a Victorian and a contemporary Christmas.

2. William J. Mitchell, *City of Bits: Space, Place, and the Infobahn* (Cambridge: MIT Press, 1995).

3. Carol Levin, "Don't Pollute, Telecommute," *PC Magazine*, 22 February 1994, 32; Jim Ludwick, "Home at the Office," *Missoulian*, 17 November 1996; Hope Lewis, "Exploring the Dark Side of Telecommuting," *Computer World*, 12 May 1997, 37; Susan J. Wells, "For Stay-Home Workers, Speed Bumps on the Telecommute," *New York Times*, 17 August 1997.

4. "A Matter of Degrees: Colorado Governor Roy Romer on the Western Governors University," *Educom Review*, January/February 1997, 18–19, 23.

5. Bill Gates, *The Road Ahead* (New York: Viking, 1995), 80.

6. Ken Auletta, "The Magic Box," *New Yorker*, 11 April 1994, 42.

7. William Gibson, *Neuromancer* (New York: Ace, 1984); Philip K. Dick, *Do Androids Dream of Electric Sheep?* (New York: Ballantine, 1996 [1968]); Gates, *The Road Ahead;* Nicholas Negroponte, *Being Digital* (New York: Vintage, 1995).

8. Aristotle, *Metaphysics*, the first sentence (980a).

9. Lee Tangedahl and Jackie Manley, "Computer Cowboys," *Montana Business Quarterly* 34 (autumn 1996): 11–12.

10. Ibid., 14.

11. Ibid.

12. See, e.g., http://pasture.ecn.purdue.edu/~mmorgan/PFI/graphic.html. 26 June 1998.

13. Tangedahl and Manley, "Computer Cowboys," 13.

14. See, e.g., the Adventure GPS Products site: http://www.gps4fun.com/gps_fun.html (10 June 1998).

15. Jennifer Cypher and Eric Higgs, "Colonizing the Imagination: Disney's Wilderness Lodge"(Vancouver: Centre for Applied Ethics). Available: http://www.ethics.ubc.ca/papers/invited/cypher-higgs.html. 26 June 1998.

16. Alan Trachtenberg, *The Incorporation of America: Culture and Society in the Gilded Age* (New York: Hill and Wang, 1982), 122.

17. Ibid., 125.

18. Bertrand Russell, "Knowledge by Acquaintance and Knowledge by Description," *Proceedings of the Aristotelian Society*, New Series 11 (1910–11): 108–28.

19. Winfried Franzen, "Die Sehnsucht nach Härte und Schwere," *Heidegger und die praktische Philosophie*, ed. Annemarie Gethmann-Siefert and Otto Pöggeler (Frankfurt: Suhrkamp, 1988), 78–92.

20. Bill McKibben, *The End of Nature* (New York: Anchor, 1989).

21. Michael Heim, *Electric Language: A Philosophical Study of Word Processing* (New Haven, Conn.: Yale University Press, 1987).

22. Gordon G. Brittan Jr., "The Secrets of the Antelope," presented at the Claremont Graduate School, 25 November 1996. Brittan's paper will rekindle *our* respect for animals as well.

23. James Welch, *Fools Crow* (New York: Viking, 1986).

24. Robert Wright, "The Evolution of Despair," *Time*, 28 August 1995, 50–57; and *The Moral Animal: Evolutionary Psychology and Everyday Life* (New York: Pantheon, 1994). Evolutionary theory underdetermines culture and morality. The further back it reaches in evolution, the less pertinent are its findings to the human condition. Nonetheless, evolutionary psychology can uncover significant boundary conditions of human well-being as Wright amply demonstrates.

25. Forrest and Flossie Galland Poe, *Life in the Rattlesnake*, ed. Mark Ratledge (Missoula, Mont.: Art Text Publication Service, 1992).

26. Public Law 96-476, 96th Congress, 19 October 1980.

27. Lawrence Haworth, *The Good City* (Bloomington: Indiana University Press, 1963); Daniel Kemmis, *The Good City and the Good Life* (Boston: Houghton, 1995).

28. Vincent Scully, "The Architecture of Community," *The New Urbanism*, ed. Peter Katz (New York: McGraw Hill, 1992), 221–30.

29. Herbert Muschamp, "Can New Urbanism Find Room for the Old?" *New York Times*, 2 June 1996.

30. Ibid., 27.

31. Ibid.

32. Jane Jacobs, *The Death and Life of Great American Cities* (New York: Vintage, 1961), 241–90.

33. *Pavarotti: Great Studio Recordings of His Central Park Program*, London Records 443 220–2.

34. Julie V. Iovine, "Tenacity in the Service of Public Culture," *New York Times*, 12 December 1995.

35. *Select Orations and Letters of Cicero*, ed. Francis W. Kelsey, 2d ed. (Boston: Allyn and Bacon, 1894), 155 (my translation).

36. Ben Sira (also called Sirach or Ecclesiasticus) 44:1 NRSV.

37. Ben Sira 44:8–9 NRSV.

38. *Matthew Paris's English History from the Year 1235 to 1273*, tr. J. A. Giles, 3 vols. (London: Bohn, 1852–54), 2:243.

39. Hans Moravec, *Mind Children: The Future of Robot and Human Intelligence* (Cambridge: Harvard University Press, 1988); Frank J. Tipler, *The Physics of Immortality: Modern Cosmology, God, and the Resurrection of the Dead* (New York: Anchor, 1995).

40. *The Ecclesiastical History of Orderic Vitalis*, ed. and tr. Marjorie Chibnall, 6 vols. (Oxford: Clarendon, 1968–80), 3:284.

41. Steven Weinberg, *Dreams of a Final Theory* (New York: Pantheon, 1992), 211–40.

42. For pictures of the balance, see Kees Boeke, *Cosmic View: The Universe in Forty Jumps* (New York: John Day, 1957); and Philip Morrison and Phylis Morrison, *Powers of Ten: A Book about the Relative Size of Things in the Universe and the Effect of Adding Another Zero* (New York: Scientific American Library, 1982).

43. Wayne C. Booth, "The Company We Keep," *Daedalus* 111 (fall 1982): 33–59.

44. Weinberg, *Dreams*, ix, 3–18.

45. From the sequence of the requiem, attributed to Thomas of Celano (1190–1260). The translation that follows is mine.

210; and geology, 172; not humanly legible, 182; impact of, on knowledge, 203–10; physical fragility of, 195–96; promise of, 3–4; rivals and replaces reality, 181–82; self-realizing, 182; social fragility of, 195–96; becomes specular, 216; structural fragility of, 196–98; both surpasses and fails to equal natural and cultural information, 219–20; about the tower of Freiburg Minster, 120; and visualization in mathematics and physics, 172

information age: characteristic mood of, 215–16. *See also* cyberspace

information relation: restored by reading, 88; terms of, 17–22; truncated in technological information, 183; truncated by writing, 47

information revolution: banality of, 215–16; claims made on behalf of, 3, 202–3

information technology: alleged to provide for immortality, 230; and business, 210–11; defined, 166–67; and devastation and loss of meaning, 230; enhances knowledge, 216; expansion and integration of, 213–14; future of, 213–16; helpful and necessary vs. distracting and dispensable, 215; and learning and teaching, 204–8; makes reality light, 216; reduces requirements of skill, 79. *See also* computers

information theory: ambiguity (equivocation) in, 185–86; amount of information according to, 131–38; and ethics, 6; initial excitement about, 132–33; limits of, 136–40; noise and ambiguity (equivocation) in, 201; rise of, 9. *See also* Shannon, Claude; Weaver, Warren

innovation: discounted by public, 215

integration: of information technology, 213–14

Intel chip: development of, 164; 8080, 158, 160

intelligence: as the capacity to restore context to written information, 88; as the capacity to retain information, 38–39; degrees of musical, 180; drops out in virtual reality, 182–83; drops out in written information, 47; requirements of, reduced by information technology, 79; needs to be strong when resolution of information (sign to content ratio) is low, 37, 180, 256n. 6; as a term of the information relation, 22

Jacobs, Jane, 225

Jefferson, Thomas, 76–77, 109, 169–70

Karsten, Harries, 256n. 15

Katz, Jon, 192

Kivy, Peter, 95

knowledge: by acquaintance and by description, 14–15; direct and indirect, 14–15; vs. having information, 14–15; enhanced by information technology, 216; intimate with natural information, 26; as affected by technological information, 203–10

Landauer, Thomas, 261n. 48

landmark: campanile of San Marco as, 10; circumscribes nearness, 219; as disclosure of cosmic or divine meaning, 20; and grids, 169–70; orders reality, 25–26, 37; tower of Freiburg Minster as, 112–13. *See also* vision quest

language: meshing with reality, 63–64, 73; as a natural sign: 45–46, 63; not structured all the way up, 62–63; synthesis of, 62–63

latitude: determined by Lewis and Clark, 170; and division into townships and sections, 76–77

learning: as apprenticeship, 203–4; and autonomy, 204; and bodily involvement, 20, 203–4, 208; through information technology, 204–8; through lecture, 204

lecture, 204
Leibniz, Gottfried Wilhelm, 33, 129,
    141–42
Leonhardt, Gustav, 94, 102
letters: as a complete and irreducible set
    of elements, 60–63; and meaning,
    61–62. See also alphabet; literacy; no-
    tation; writing
Lewenstein, Bruce, 208–9
Lewis, Meriwether, 168–70
library: need for, today, 212
lightness of reality: brighter and less
    heavy, 216; due to information tech-
    nology, 216
Linear B, 20–21
linear measures: and information, 74
literacy: and liberation, 53; as a skill and
    a cultural force, 47
logical reasoning, 151–52
logic gates, 145–49, 152–54
logographic writing systems, 21
London, 199–200
longitude: determination of, 78–79; and
    division into townships and sections,
    76–77; information for, gathered by
    Lewis and Clark, 170; and prime me-
    ridian, 77

Machlup, Fritz, 9, 235n. 2 (chap. 1)
MacKay, Donald, 11, 132, 135, 256n. 4
Maclean, Norman, 89, 90–91
Mandel, Tom, 191–92
maps: in ancient Greece, 75; making of,
    168–70; and printing, 82; and survey-
    ability, 78, 168–69
mathematics: visualized through techno-
    logical information, 172–74. See also
    algebra; arithmetic; geometry
Mauchly, John, 143
McKibben, Bill, 26–27, 178, 219
McLeod, Milo, 30
meaning: as content of information, 18;
    decline of, 10, 15, 33, 202; —, and
    weak readings of history, 232; emer-
    gence of, 20, 29; place of, in informa-
    tion relation, 18; remembrance and

redemption of, 229, ruination and
    evaporation of, 228–29
measures: of information, 131–38, 251n.
    18; origin of linear, 74; ultimate, of
    reality, 126–27
Melton, Douglas, 30
memory: artificial vs. natural, 241n. 3;
    cultivated and supported in oral cul-
    ture, 38–40; as information space,
    168; vs. writing, 47–48, 50–51
Memory Theater, 175
message: as content or meaning of infor-
    mation, 18; distinguished from sig-
    nal, 179. See also information, con-
    tent of; meaning
Mitchell, William, 15–16, 187, 256n. 16
money: as a means of producing infor-
    mation, 83–84
Montana: cyberspace in, 216–18; and its
    grid of longitudes and latitudes,
    76–77
MOO (Multi-user Domain, Object Ori-
    ented), 188–91
Moore, Gordon, 144
Moore's Law, 144
moral and material dimensions of real-
    ity: inversion of, 220–21
moral and material prosperity, 1, 57
Moriarty, Gene, 237n. 1
Moulthrop, Stuart, 210
Mozart, Wolfgang Amadeus, 57–59, 97,
    103, 121
MUD (Multi-user Domain), 188–91
Muschamp, Herbert, 225
music: how digitized on a CD, 197; nota-
    tion of, 83; recorded, 103; as struc-
    ture, 93–96; structures time, 103–4
music performance: and communal prac-
    tices, 103–4; comprehension of, 95–
    96, 100; crucial to music, 98; as the
    realization of information, 93, 100;
    resolves the ambiguities of the score,
    99–100; richness of its structure, 96;
    variety of styles of, 95, 99, 100
music score: and historical contingency,
    99–103; performance of, 99–100; as

pure structure, 94–96; realization of, 94–96; as signified structure, 93

nearness: circumscribed by landmark, 219; and farness, 15; as focal area of presence, 25; and information, 15. *See also* landmark; presence; reality, eloquence of
Neumann, Werner, 94
Newton, Isaac, 72. *See also* physics, Newtonian
New Urbanism, 224–25
New York: layout of, 77; festivity in, 226–27
Nietzsche, Friedrich, 54
noise: cultural, 201–3; in education, 203; in fiction, 210; in information theory, 201; in research, 208–9
Norman England: and the spread of literacy, 50–51
notation: binary, 129–31; and contingency, 98; musical, development of, 83; music CDs and Nelson Goodman's criteria of, 258n. 10; and purity of structure, 98. *See also* alphabet; letters
Noyce, Robert, 164
numeracy. *See* counting

Oersted, Hans Christian, 126

Pavarotti, Luciano, 226
People of the Book, 31
performance. *See* music performance
Perseus, 209
person: as ambiguous sign, 231–33; drops out in written information, 47; mind of, as distance learner, 206, 259 n. 32; as a term of the information relation, 18
personal computer: and personality of learner, 259n. 32
perspicuity: as benefit of natural information, 1–2; expanded into surveyability, 78, 168–69; superseded by transparency, 168–69

physics: and lawful structure, 99; Newtonian, 72, 126–27; and structural cosmic closure, 32–33, 231–32
Plato: his appreciation of the alphabet, 59–60; on constructing reality from the regular polyhedra, 61–62, 125; his critique of Homeric orality, 59–60; and language as a conventional system, 63–64; and literacy vs. community, 52; and his misgivings about writing, 47–49; on reading and context, 88; and writing as material, 53–54.
playing. *See* music performance
polyhedra, regular: as the Platonic building blocks of reality, 61–62
possibility space: constructed by reading, 85, 92, 93; and creativity, 138–39
postmodern divide, 202
postmodern lightness of reality, 220
presence: and eloquence, 29; and reference, 1, 17–18; imperilled by transparency, 176–77; and information, 14–15; unencumbered by reference, 25, 49. *See also* landmark; nearness; reality, eloquence of
printing: and astronomical and mathematical tables, 82; as grid imposed on reality, 80; renders information precise and universal, 81, 82–83; and maps, 82
proportionalism, 114–16
prosperity: advanced by technological information, 177–78; as benefit of cultural information, 1–2; material and moral, 1, 57
Ptolemy, 75, 82
public has discounted innovation, 215
pun, 19
Pythagoreanism: and purity of structure, 96–97
Pythagorean theorem: and circle in analytic geometry, 69–71; and having a computer program find Pythagorean triples, 161–62; in the construction of the agni, 65–68; in Euclid, 68–69